高等职业教育“十四五”系列教材
高等职业教育土建类专业“互联网 +”数字化创新教材

建筑英语

English for Building Construction

戴明元　邓冬至　主　编
罗忠明　吕　颖　王建忠　副主编
袁　俊　主　审

中国建筑工业出版社

图书在版编目（CIP）数据

建筑英语 = English for Building Construction / 戴明元，邓冬至主编；罗忠明，吕颖，王建忠副主编；袁俊主审 .—北京：中国建筑工业出版社，2022.9（2025.1重印）
高等职业教育“十四五”系列教材 高等职业教育土建类专业“互联网 +”数字化创新教材
ISBN 978-7-112-27531-1

Ⅰ.①建… Ⅱ.①戴… ②邓… ③罗… ④吕… ⑤王…⑥袁… Ⅲ.①建筑—英语—高等职业教育—教材
Ⅳ.① TU

中国版本图书馆 CIP 数据核字（2022）第 103744 号

本书共 10 个单元，以建筑施工为主线，内容涉及现场管理、基础施工、框架结构、屋面工程、围护结构、室外工程、水电安装、室内装饰、园林绿化、中国古今优秀建筑文化介绍等。每个单元围绕一个主题，由两部分组成，包括工作场景会话、专业知识概述、阅读理解。每部分配有插图、专项练习，突出职业能力培养，侧重建筑施工环境下的语言交际能力的培养。本书还配有工作手册、富媒体资源、导学视频等。

本书可供高等职业院校、应用型本科院校及成人院校建筑类专业的学生使用，也可供建筑类专业国际学生及涉外建筑工程管理人员使用。

为方便教师授课，本教材作者自制免费课件，索取方式为：

1. 邮箱 jckj@cabp.com.cn；2. 电话（010）58337285；3. 建工书院 http://edu.cabplink.com。

责任编辑：李天虹　李　阳
责任校对：张惠雯

高等职业教育“十四五”系列教材
高等职业教育土建类专业“互联网 +”数字化创新教材
建筑英语
English for Building Construction
戴明元　邓冬至　主　编
罗忠明　吕　颖　王建忠　副主编
袁　俊　主　审
*
中国建筑工业出版社出版、发行（北京海淀三里河路 9 号）
各地新华书店、建筑书店经销
北京雅盈中佳图文设计公司制版
建工社（河北）印刷有限公司印刷
*
开本：787 毫米 × 1092 毫米　1/16　印张：$12^3/_4$　字数：301 千字
2022 年 9 月第一版　2025 年 1 月第二次印刷
定价：38.00 元（赠教师课件）
ISBN 978-7-112-27531-1
（39677）

前言

《建筑英语》是一本供高等职业院校、应用型本科院校建筑类专业学生及土建类国际学生教育、涉外建筑企业技术人员培训使用的教材。本教材以先进的英语教育理念为指导，突破行业英语教材编写体例，融合英语和建筑知识，着重培养学生国际工程基本技能和建筑施工环境下英语应用能力，促进学生适应建筑国际化发展。

本教材的主要特色体现在：

一、双元建设：本教材从规划、素材选择、编写等环节注重产教融合、校企合作。四川华西海外投资建设有限公司与四川建筑职业技术学院建立了长期的校企合作关系，是学院在海外的重要实训实习基地。公司积极支持《建筑英语》编写工作，总经理王建忠担任副主编，使教材规划与编写充分发挥行业作业。

二、课程思政：本教材把语言培训、技能传授与中国优秀建筑文化有机结合，培养学生政治认同、国家意识、文化自信、爱国情怀，突出"立德树人"根本任务。

三、职业特色：本教材对接国际先进职业教育理念，吸收新技术、新工艺、新规范，介绍绿色环保、健康住宅。注重理论与实践的统一，实践中提升语言应用能力，以一线施工活动、工作任务、工程案例等为载体组织教学单元，力求为学生展示一个完整的施工过程，突出职业教育特色。

四、富媒体资源：本教材配有音频诵读、导学视频、思维导图、微课、课件、教学参考书等资源，便于学生学习、教师教学。

五、工作手册：本教材编制工作手册指导学生实践。课前任务、课中任务及课后任务充分展示教学内容的实用性、时代性和实践性，培养学生思维能力、英语交际能力和建筑知识应用能力。

本教材特色还在于其结构安排合理，教学内容精心挑选。每单元围绕一个主题进行循序渐进的语言技能和建筑知识训练，提高学生专业英语的表达能力和阅读能力，增强学生对建筑知识的理解。每单元配有与内容相关的插图，既有利于学生理解建筑知识，又为学生语言实践提供情景。

本教材的编写遵循"凡编必审"原则，由四川建筑职业技术学院应用英语专业（国际工程翻译方向）建设负责人袁俊教授担任主审，对其内容的思想性、科学性、适宜性，语言的规范性进行了审核。

本教材共 10 个单元，以建筑施工为主线，内容涉及现场管理、基础施工、框架结构、屋面工程、围护结构、室外工程、水电安装、室内装饰、园林绿化等。每单元安排如下：

Learning Objectives 列出包括本单元主题层面、语言技能层面及建筑知识层面的学习目标。

Part One

此部分包括两个方面，Section A Workplace Conversation 是本教材的特色之一，以现场施工常见问题为素材，着重培养学生解决问题的能力、建筑施工环境下英语交际能力和英语写作能力；Section B 是对本单元专业知识的概述，使学生对相关的专业知识有基本的了解。该部分的练习除专业词汇和阅读理解练习外，还设计了“识图会话”练习，以提升学生的专业知识理解能力和语言应用能力。

Part Two

此部分围绕本单元的主题选材，是 Part One 的补充和加强。包括课文、专业词汇和词组 (Professional Words and Expressions)、课后练习 (Exercises)。课后练习包括阅读理解练习 (Reading Comprehension)、翻译 (Translation)、小组活动（Group Activities）等。Group Activities 紧密结合本单元主题，旨在使学生通过小组活动对主题有更多的了解，同时加强语言实践能力的训练，扩展学生的建筑知识，增强学生的语言表达及交际能力。

The Pride of Chinese Architecture 介绍中国古今优秀建筑文化，提升学生阅读能力，培养学生爱国情怀、文化自信。

本教材由四川建筑职业技术学院长期从事职业教育的一线教师和四川华西海外投资建设有限公司具有多年海外工程建设经验的管理者共同编写，戴明元教授、邓冬至教授担任主编，罗忠明教授、吕颖副教授、王建忠总经理担任副主编，张义琢高级工程师、王娇娇副教授、文阳讲师参与编写。其中戴明元编写第 1 章和第 10 章，邓冬至编写第 2 章和第 7 章，罗忠明编写第 3 章，吕颖编写第 4 章和第 6 章，王建忠编写第 5 章，王姣姣编写第 8 章，文阳编写第 9 章。从事国际工程施工及管理的高级工程师张义琢老师对教材内容的专业性、建筑术语的准确性及中英文表达习惯做了校对和修订，绘制了部分插图。

由于编者专业水平有限，书中疏漏和不妥之处在所难免，敬请各位读者不吝指正。

Guide Learning of *English for Building Construction*

★ I. Why shall we learn *English for Building Construction*?

In 2013, General Secretary Xi Jinping put forward the Belt and Road Initiative, and in 2014, Premier Li Keqiang put forward the concept of "international production capacity cooperation" , which promoted the integration of infrastructure construction and industrial structure upgrading and promoted the outward development of the construction industry. However, the "going global" construction industry is in serious shortage of site managers and construction technicians who understand building construction and English. To meet the needs of "going global" construction industry for professional personnel, we have compiled the *English for Building Construction*.

English for Building Construction is a textbook for architectural students in Higher Vocational Colleges and Applied Universities. The compilation of this textbook is based on the research of English for special purpose, draws lessons from the advanced English education concept, combines the characteristics of the international development of the construction industry, breaks through the original compilation style of English for special purpose, integrates English and architectural knowledge, and focuses on cultivating students' English application ability in the construction environment.

★ II. What shall we learn from *English for Building Construction*?

English for Building Construction takes building construction as the main line and covers on-site management, foundation construction, frame structure, roof engineering, enclosure structure, external works, hydropower installation, internal decoration, landscaping, etc. Each unit revolves around a theme and consists of two parts, including workplace conversation, professional knowledge overview and reading comprehension. Each part is assigned with illustrations and special exercises, highlighting the cultivation of professional ability and focusing on the cultivation of language communication ability in the construction environment. By studying ***English for Building Construction***, we shall not only learn architectural knowledge, but also improve our English application ability.

English for Building Construction also introduces excellent ancient and modern Chinese architecture to enhance our understanding of China's excellent architectural culture.

★ Ⅲ. How shall we learn *English for Building Construction?*

English for Building Construction is a practical course. To learn the course well, we suggest that:

1. Understand the learning objectives of each unit and be prepared for learning.

2. Get background information as much as possible before learning each unit.

3. Prepare well before class, read the text and learn the professional words and expressions by heart.

4. Listen carefully and make notes while learning in the class.

5. Apply what you have learned and actively participate in class discussions.

6. Finish the homework as required.

7. Practice in listening, speaking, reading and writing persistently.

★ Ⅳ. What shall we get from *English for Building Construction*?

English for Building Construction is to cultivate technical and skilled talents who can speak English, understand construction technology and meet the needs of foreign-related projects. Therefore, having learned this course, we are able to:

1. Improve the English application ability and English communication ability under the construction environment, so as to lay a good foundation for career development.

2. Acquire construction knowledge, management knowledge and on-site construction experience to prepare for international building construction.

3. Expand international vision, learn business negotiation, and serve the construction of "the Belt and Road" and international production capacity cooperation.

4. Better understand China's excellent architectural culture and enhance cultural self-con fidence.

English for Building Construction is an important tool to realize "speaking English, understanding engineering, knowing management and being good at communication" .

CONTENTS

Unit One

Concept of Building and Construction Engineering

Learning Objectives

After learning this unit, you will be able to

- write a site report according to the workplace conversation;
- identify drawings and documents used on site;
- list the constituent parts of a building;
- explain building elements and their function;
- use the correct terminology in the description of building;
- understand Chinese excellent architectural culture.

扫一扫，听录音

Part One

Section A Workplace Conversation

Professional Words and Expressions

contractor [kən'træktə (r)]	*n.*	承包商
client ['klaɪənt]	*n.*	雇主
commencement [kə'mensmənt]	*n.*	开始
deliver [dɪ'lɪvə (r)]	*v.*	提交
architect ['ɑ:kɪtekt]	*n.*	建筑师
instruction [ɪn'strʌkʃn]	*n.*	指令
elevation [ˌelɪ'veɪʃn]	*n.*	立面图
section ['sekʃn]	*n.*	剖面图
specification [ˌspesɪfɪ'keɪʃn]	*n.*	技术说明书
professional [prə'feʃənl]	*adj.*	专业的
engineering [ˌendʒɪ'nɪərɪŋ]	*n.*	工程
letter of acceptance		中标函
performance security		履约担保
access to the site		进入现场
take over		接收
quality control		质量控制

In the building industry, the usual arrangements are that the contractor constructs the works in accordance with a design provided by the employer (client),or his representative, the Engineer. The contractor gets paid for the work he performs and the employer gets the work he is paying for. Before the commencement of the work, the contractor is required to submit several documents.

★ **I. Listen to the conversation and fill in the blanks with what you hear.**

(George – Client; Lucas – Contractor; Lucy – Architect)

George: Hi, Mr. Lucas, congratulations! Your company is awarded the 1)____________ for the project. This is the letter of acceptance.

Lucas: Oh, thank you, George! We are so pleased to 2)____________.

George: Before you commence the work, you will have to deliver the performance security to us, and 3)____________.

Lucas: Yes, we will.

George: And you will have to submit a detailed program which indicates the time for 4)____________.

Lucas: Sure, we will do that. May I ask when we will have the access to the site?

George: The Architect will give you instruction for the commencement date after we have received your performance security.

Lucas: Oh, that will be all right. When can we get all the drawings?

George: The architect said the plan, the elevation and the 5)_____________ drawings were all ready, as well as specifications. You can get them from her office.

Lucas: OK, thank you. I'll talk to the Architect.

George Hope you will have a good cooperation with the Engineers and I can 6) _____________ a perfect work.

Lucas: We are the professional engineering company. 7)_____________ is our first priority in construction. We are sure to complete the work in accordance with the Engineers' instruction, and 8)____ _____________.

★ II. Lucas writes a report according to the conversation above.

This morning, the client told me that our company was awarded the contract of the project. He issued the letter of ____
____
____
____
____

扫一扫，听录音

Section B People Who Work on Site, Drawings and Documents

Communication is an essential part of everyday life. However, the ability to communicate is often taken for granted. Can you imagine how frustrating it would be if you were unable to communicate with others? If so, no matter how hard you tried, you just could not get your point across.

Good communication is the ability to make others understand what you are trying to communicate.

Through your work in the building and construction industry you will come in contact with lots of people from many different areas. It is likely that each of these groups will use certain terminology and jargon specific to their area and it is also likely that this language will vary from country to country.

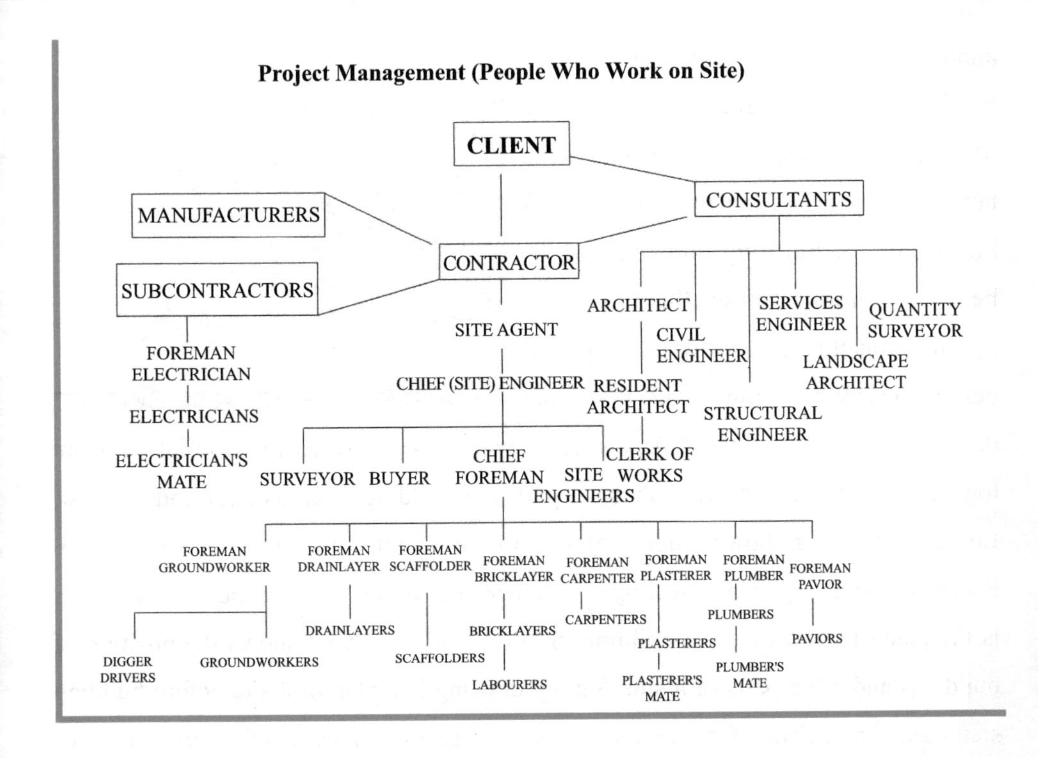

This part aims to provide you with a general knowledge of people who work on site that you are going to get in contact with in building industry, drawings and documents that are available on site.

Drawings and Documents

Building drawings are diagrams of what a building will look like when it is finished. They are prepared by architect, engineers or drafters, usually for a client, and include all the information necessary for the construction of a building. Builders use

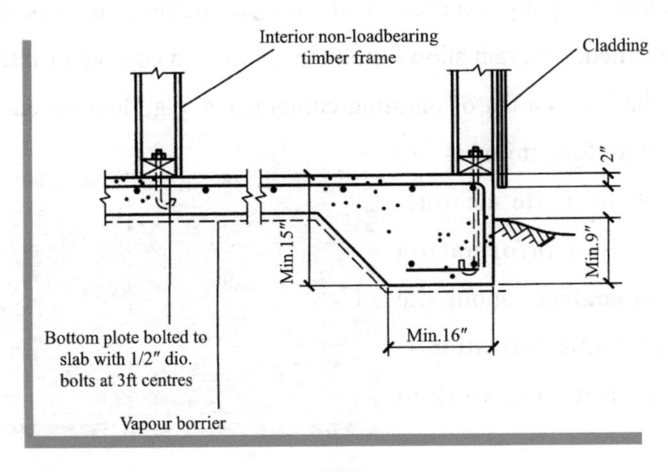

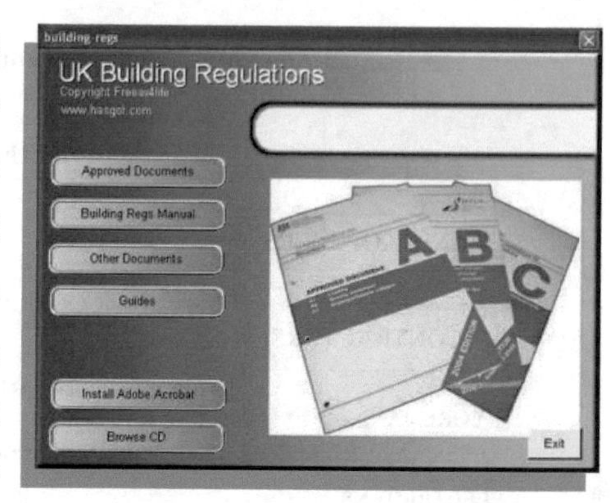

annotated (labeled) drawings or sketches to show in close detail the assembly of building material or components. It takes many years of practice and a sound knowledge of building construction to be able to produce these illustrations. It is important that you show the required details clearly and simply and that your drawings are fully annotated. Assembly drawing shows how part of a building is put together. Component drawing shows a part of a building, such as door and window. Location drawing shows where parts and components of a building are located. Perspective drawing gives a feeling of distance and solidity to the building. Floor plan is the plan of one floor of a building. Site plan or site layout shows the position of buildings and other parts of a site. Survey drawing is a plan of a site before building starts, showing existing features and levels. Elevation is a drawing of the front, back or side of a building while section is a representation of a part of the building looked at from the side.

There are many written documents you may have to read when you are working on a building site. Bill of quantities lists materials and work required to build a building. Daywork sheet records the work done which is in addition to the contract work. The form of contract is the list of terms and conditions which apply to a contract between a client and a contractor. Insurance policy is a document which lists the terms and conditions of an insurance contract, while insurance premium is sum of money paid for insurance cover, so that the damage to a building would be compensated by the insurance company. Licence means official permit which is issued by the authority concerned. Program shows the time in which a contractor intends to build a building. Schedule is a list of building components, e.g. door schedule, schedule of sanitary ware. Site minutes are written records of a site meeting. Soil report shows information prepared by an engineer about site ground conditions.Specification is the written description of work to be done.

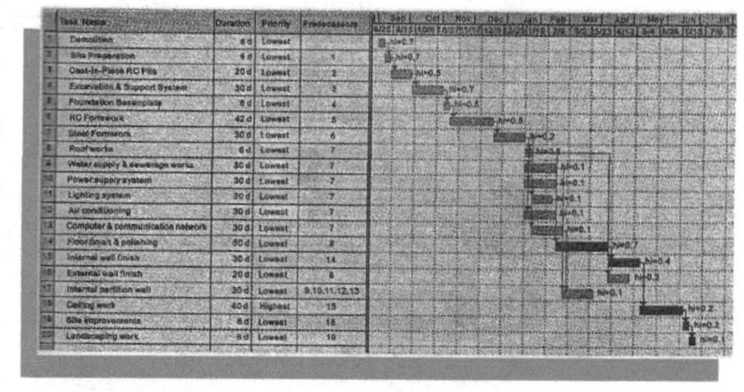

Professional Words and Expressions

consultant [kən'sʌltənt]	*n.*	（顾问）工程师，监理工程师
manufacturer [ˌmænju'fæktʃərə(r)]	*n.*	制造商
subcontractor [ˌsʌbkən'træktə(r)]	*n.*	分包商
civil ['sɪvl]	*adj.*	土木的
structural ['strʌktʃərəl]	*adj.*	结构的
service ['sɜ:vɪs]	*n.*	设备
landscape ['lændskeɪp]	*n.*	园林，景观
electrician [ɪˌlek'trɪʃn]	*n.*	电工
buyer ['baɪə(r)]	*n.*	采购员
groundworker ['graʊndwɜ:kə]	*n.*	挖土方的工人，铺路工
digger ['dɪgə(r)]	*n.*	挖掘者，挖掘机
drainlayer [drein'leiə]	*n.*	铺设排水管的工人，排水管铺设机
scaffolder	*n.*	架子工
bricklayer ['brɪkleɪə(r)]	*n.*	砖工
carpenter ['kɑ:pəntə(r)]	*n.*	木工
plasterer ['plɑ:stərə(r)]	*n.*	抹灰工
plumber ['plʌmə(r)]	*n.*	管工，水暖工
pavior ['peɪvjə]	*n.*	铺砌工人，石匠
drawing ['drɔ:ɪŋ]	*n.*	图纸
diagram ['daɪəgræm]	*n.*	（示意）图，计算图表
drafter ['drɑ:ftə]	*n.*	绘图员
annotate ['ænəteɪt]	*v.*	注释，注解
labeled ['leɪbld]	*adj.*	带有标记的
sketch [sketʃ]	*n.*	草图，设计图
assembly [ə'semblɪ]	*n.*	装配，组装；装配图
component [kəm'pəʊnənt]	*n.*	零（部）件，构件
detail ['di:teɪl]	*n.*	详图，大样
perspective [pə'spektɪv]	*n.*	透视图，远景
layout ['leɪaʊt]	*n.*	布置，定位（线），放样
survey ['sɜ:veɪ]	*v.*	测量
premium ['pri:miəm]	*n.*	保险费
resident architect		驻现场建筑师

clerk of works	现场监工员，工程代表
quantity surveyor	估算师，计量员
site agent	现场经理，项目经理

Exercises

★ Ⅰ. Answer the following questions.

1. What is good communication?
2. Whom will you come in contact with in the building and construction industry?
3. What are the relations among the client, the consultants and the contractor?
4. What are the relations among the contractor, the subcontractor and the manufacture?
5. List as many drawings and documents as possible used on the site.

★ Ⅱ. Match the following words or phrases with the correct Chinese.

1. bill of quantity	a. 文件
2. client	b. 立面图
3. elevation	c. 估算师
4. layout	d. 雇主
5. quantity surveyor	e. 土壤报告
6. floor plan	f. 工程量清单
7. subcontractor	g. 楼层平面图
8. letter of acceptance	h. 平面布置
9. soil report	i. 分包商
10. document	j. 中标函

★ Ⅲ. Read the following drawings.

1. Find a proper word or phrase to describe each of the drawings.

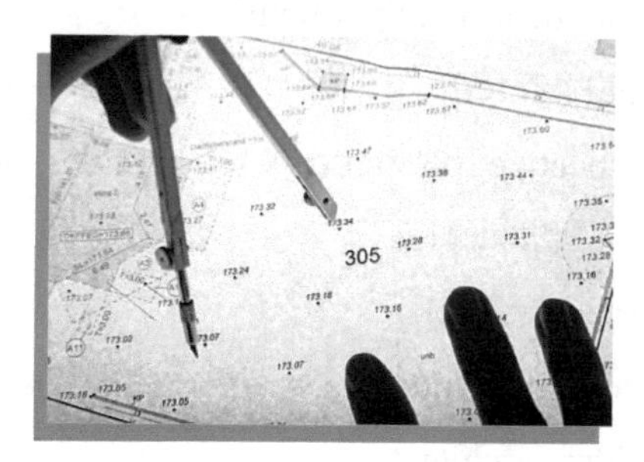

1) ______________________________　　2) ______________________________

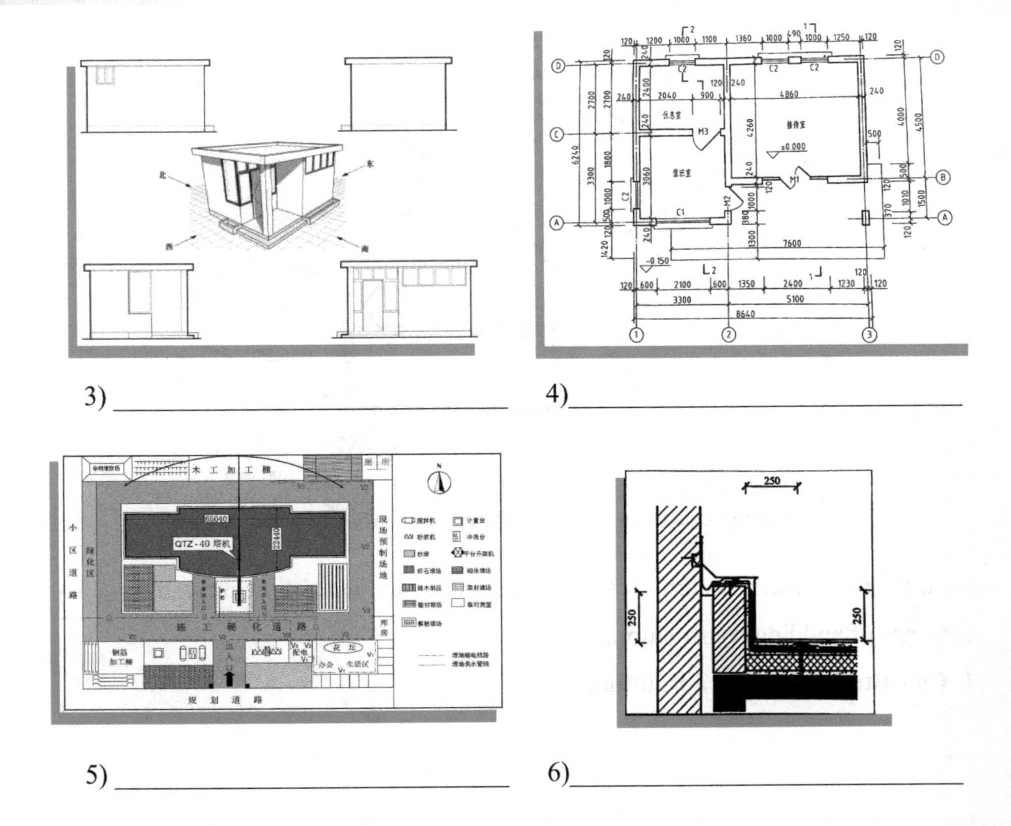

3) ______________________ 4) ______________________

5) ______________________ 6) ______________________

2. Read the drawings again and tell your partners what the function is for each drawing.

Part Two

扫一扫，听录音

Concept of Building Construction

Building Construction involves many trades, operations, products and professions. It deals with the design of the fabric of the buildings and the manner in which it is put together. Therefore, building construction requires the understanding of sciences of materials and structures, environmental sciences, and building economics. Today, as the world's energy sources are deleting and the cost of energy rises, it is becoming even more important to design and construct energy efficient housing.

A building is an enclosure for the benefit of human habitation, work and recreation. It is also an enclosure against cold, heat, wind, rain to give a comfortable

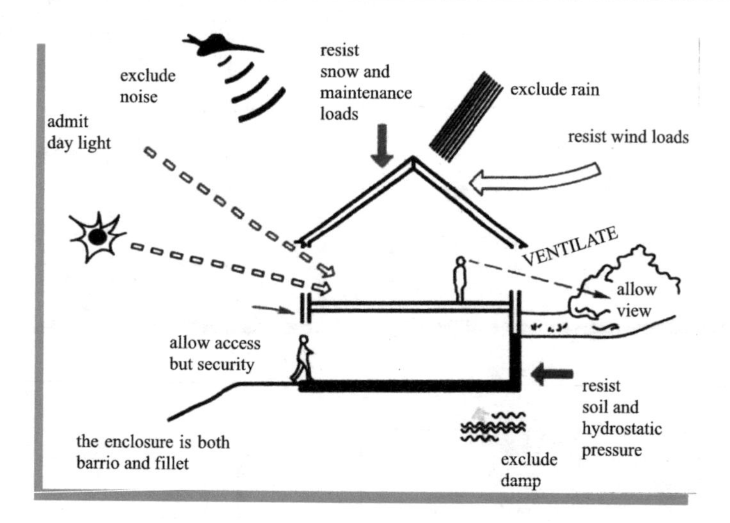

internal environment, security for life and possessions. The followings are some basic concepts of building construction:

1. Constituent Parts of a Building

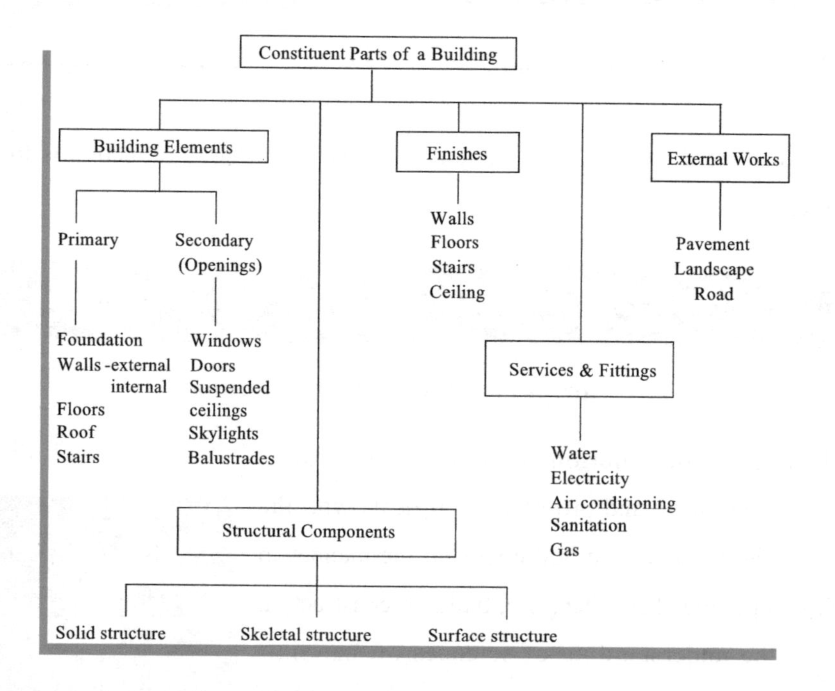

solid structure

skeletal structure

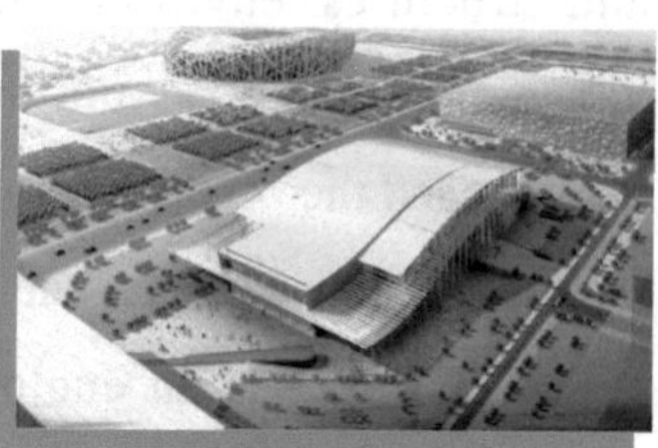
surface structure

2. Building Elements

EXAMPLES	FUNCTIONS AND TYPES
Foundation	Services to spread the loads of the building to the underlying soil e.g. strip, pad, piles, raft
External walls	Formation of the external envelop of the building e.g. load-bearing and non-bearing walls
Internal walls	To divide the space enclosed by the external walls into rooms e.g. bricks, blocks, timber, metal
Floors	To divide space enclosed by the external walls and roof horizontally to increase the usable space e.g. ground floor, upper floors
Stairs	1) for access from floor to floor 2) for escape during a fire e.g. materials, configuration
Roofs	To protect the building from the weathering elements e.g. pitched roof, flat roof
Doors	A screen to seal an opening into a building, or between rooms within a building e.g. materials (construction), methods of opening
Windows	1)to let light and sun into a building 2) to provide a view of what is inside or outside e.g. materials, methods of opening

Professional Words and Expressions

trade [treɪd]	*n.*	工种，行业，专业
fabric ['fæbrɪk]	*n.*	结构，骨架
constituent [kən'stɪtjuənt]	*n.*	构成，组成部分
finish ['fɪnɪʃ]	*n.*	装修
external [eks'tə:nl]	*adj.*	外面的
internal [in'tə:nəl]	*adj.*	内部的
suspended [sə'spendɪd]	*adj.*	悬挂的，悬挑的，悬空的
skylight ['skaɪlaɪt]	*n.*	天窗
balustrade [ˌbælə'streɪd]	*n.*	栏杆（柱），栏杆扶手
pavement ['peɪvmənt]	*n.*	路面，硬化区，（园林）小路
fitting ['fɪtɪŋ]	*n.*	配件，家（灯）具，器具
sanitation [ˌsænɪ'teɪʃn]	*n.*	卫生设备（排水设备），下水道设备

structure ['strʌktʃə (r)]	*n.*	结构，构筑物
skeletal ['skelətl]	*n.*	框架的，骨架的
surface ['sɜ:fɪs]	*adj.*	表层的
load [ləʊd]	*n.*	荷载
strip [strɪp]	*n.*	条形基础；长条，板条
pad [pæd]	*n.*	独立基础
pile [paɪl]	*n.*	桩基础
raft [rɑ:ft]	*n.*	筏形基础，排基
block [blɒk]	*n.*	砌块
timber ['tɪmbə (r)]	*n.*	木材
configuration [kənˌfɪgə'reɪʃn]	*n.*	外形，构造（形式）
energy efficient		能源高效
pitched roof		坡屋顶
flat roof		平屋面

Exercises

★ Ⅰ. Reading Comprehension

1. What does building construction involve?

2. Describe briefly the building construction requirements.

3. Why is it becoming even more important to design and construct energy efficient housing?

4. Explain the finishes in building construction.

5. According to your understanding, explain briefly the structure in building construction.

★ Ⅱ. Translation

A. Translate the following sentences into English.

1. 在建筑施工中，你会接触到来自不同地方的各种人员。

2. 你给出的大样图要简洁明了，图纸的注释要充分，这很重要。

3. 从事建筑施工需要懂得材料学、结构、环境学及建筑经济学。

4. 雇主任命工程师，工程师应履行合同中指派给他的任务。

5. 解决能源短缺矛盾的较有效方法是进行建筑节能控制。

B. Translate the following paragraph(s) into Chinese.

As the world's energy sources are deleting and the cost of energy rises, it is becoming even more important to design and construct energy efficient housing. An energy efficient house is defined as a house which results in greater comfort and lower running costs. Not only does a well designed house save money but it is also environmentally friendly.

Important decisions regarding the needs of an energy efficient house need to be made early in the design stages to receive the most benefit from the house's surrounding environment. The proper use of windows, construction materials, insulation and natural ventilation all play a major part in the success of an energy efficient design.

The principle aim of low energy housing is to use energy carefully to achieve a reasonable level of comfort and obtain an optimum balance between the cost of construction and the cost of heating and cooling a house. This means that a low energy house should:

· have minimum heat loss and maximum solar gain in winter;

· have minimum heat gain and maximum natural cooling in summer.

This is called the summer/winter compromise.

★ Ⅲ. Group Activities

1. What is energy efficiency and how to design energy efficient house?

2. How does the environment effect on human comfort?

____

____

____

____

____

The Pride of Chinese Architecture

扫一扫，听录音

Bird's Nest: The National Stadium, widely known as the Bird's Nest because of its unique design, has become a nationwide sensation. It is situated in the Beijing Olympic Green. During the Beijing 2008 Olympic Games, the opening ceremony, closing ceremony, track and field meets, and men's football final were held here. It was also the host to the opening and closing ceremonies for the Beijing 2022 Winter Olympic Games. It is the world's largest steel structure and the most complex stadium ever constructed. It has been a landmark and a longstanding symbol of the Olympics, and is "one of the key engineering marvels in the world today."

Self-assessment

Now I know

· people who work on site include ____;

· drawings which are required on site include ____;

· documents which are required on site include ____;

· building construction can be classified into ____________________;

· building elements usually include ____.

Now I can

☐ use the correct terminology in the description of building;

☐ write a site report according to the workplace conversation;

☐ describe drawings and documents used on site;

☐ distinguish the constituent parts of a building;

☐ explain building elements and their functions.

Unit Two

Preliminary Site Work

Learning Objectives

After learning this unit, you will be able to

- write a site report according to the workplace conversation;
- understand the meaning of pre-construction activities;
- know the importance of site investigation;
- know what should be included in site analysis;
- use the correct terminology in the description of preliminary site work;
- understand Chinese excellent architectural culture.

扫一扫，听录音

Part One

Section A Workplace Conversation

Professional Words and Expressions

stockpile ['stɒkpaɪl]	*v.*	堆放
topsoil ['tɒpsɔɪl]	*n.*	表层土
courtyard ['kɔ:tjɑ:d]	*n.*	院子
scaffold ['skæfəʊld]	*n.*	脚手架
platform ['plætfɔ:m]	*n*	平台
storey ['stɔ:rɪ]	*n.*	楼层
hut [hʌt]	*n.*	棚屋，临时用房
accessible [ək'sesəbl]	*adj.*	可接受的
foundation [faun'deiʃən]	*n.*	地基
watertight ['wɔ:tətaɪt]	*adj.*	水密的，不漏水的
set out		放样
site plan		现场平面图
site engineer		工地工程师，施工员

How to Build on a Crowded Site

The first thing a contractor looks for on a site plan is working space. If there is not enough room in which to work, he may have to think carefully about which building should be constructed first.

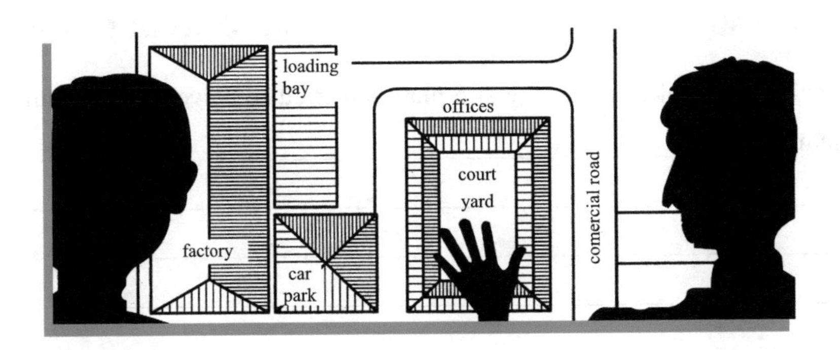

★ I. Listen to the conversation and fill in the blanks with what you hear:

(Peter — Site Agent; Kevin — Site Engineer)

Peter: Just look at this 1)______________. Kevin, we're going to have problems here, you know.

Kevin: Hmm-yes, I can see that. There isn't really enough room to work, is there?

Peter: Well, no. I can't see where we'll be able to 2)____________________, for a start.

Kevin: No. And I don't know where we're going to put out site offices either - Well, perhaps we could try putting them inside the courtyard. Do you think it's big enough?

Peter: Well, let's see - how much space is there? -Hmm, only about a hundred 3)____________________. Well, that won't be big enough, will it?

Kevin: Well, no-but I suppose we could always use a scaffold platform and have two - storey offices.

Peter: Oh no, they're so inconvenient. And anyway, 4)____________ would be in the way there, really. But listen, we can do it another way. I think, the factory is the last 5)______________________, isn't it?

Kevin: Yes, I think so.

Peter: Then it would be a good idea to set that out first. We could get 6)____________ in there quickly, couldn't we?

Kevin: Oh yes. And we could put the site huts up there while all the other 7)_____________ are being laid.

Peter: That's right. And after that I suggest we get on with the office block straight

away. If 8)____,
we'll be able to use them as our own offices, you see. Won't we?

★ Ⅱ. Peter writes a site report according to the conversation above.

Kevin and I talked about the site plan and tried to solve the problem in site arrangement. ____
____
____
____
____

扫一扫，听录音

Section B Concept of Preliminary Site Work

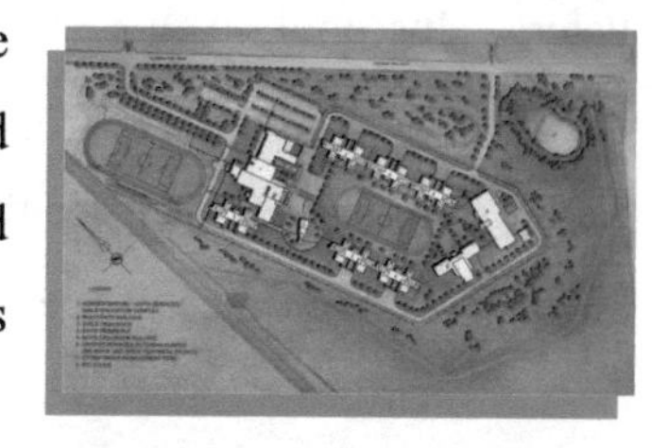

Before any construction of a building can take place, a series of operations are to be carried out, termed as pre-construction activities. Plans, financing and building permits are included in this stage, as well as site investigation and preliminary site works.

Site investigation: Site investigation is required to determine the factors which may affect the design and the construction processes of the proposed works. There are two types of site investigation, above ground investigation and below ground investigation. The above ground investigation includes investigating the site, such as obstruction, existing building and topography, the surrounding area, such as neighboring properties, and the facilities near the site, e.g. nearest town for facilities & labour. The below ground investigation includes the soil investigation which is to determine strength of soil and to detect presence of causes, mines and harmful materials. It also investigates the existing underground services such as water pipes, sewages pipes, electrical cables, telephone lines, and gas pipes underground. The purpose of the investigation is to achieve better design, economical preliminary work and better construction method.

Preliminary site works: Preliminary site works is to be done before the actual construction work begins on site, which usually involves site clearance and site layout.

Site clearance includes demolition of existing structure, removal of trees, shrubs and topsoil, and diversion of existing services. Site layout has to consider the access for entry and exit the site, storage facilities for materials, workshops or fabrication areas for formwork and reinforcement, mechanical plant such as tower crane and mixing plant for concrete production, site offices for administrative function, welfare & sanitary facilities such as temporary living quarters, site canteen, toilets / bathrooms for the staff and workers, security & protection for preventing theft of materials and injury to members of public, and temporary services such as electricity, water and telephone lines.

Professional Words and Expressions

obstruction [əb'strʌkʃn]	*n.*	障碍物
topography [tə'pɒgrəfi]	*n.*	地势，地貌
sewage ['su:ɪdʒ]	*n.*	下水道
layout ['leɪaʊt]	*n.*	布局
demolition [ˌdemə'lɪʃn]	*n.*	拆除
fabrication [ˌfæbrɪ'keɪʃn]	*n.*	装配
formwork ['fɔ:mwɜ:k]	*n.*	模板
reinforcement [ˌri:ɪn'fɔ:smənt]	*n.*	钢筋，加固
concrete ['kɒŋkri:t]	*n.*	混凝土
sanitary ['sænətri]	*adj.*	卫生的，清洁的
obstruction [əb'strʌkʃn]	*n.*	障碍物
tower crane		塔式起重机
mixing plant		搅拌厂
site office		工地办公室

Exercises

★ I. Answer the following questions.

1. What do the pre-construction activities include?
2. Why is the site investigation required?
3. What does the below ground investigation include?

4. When is the preliminary site works to be done?

5. What does the site layout have to consider?

★ Ⅱ. Match the following words or phrases with the correct Chinese.

1. demolition	a. 障碍物；阻碍，障碍
2. sewage	b. 布置，布局；定位
3. obstruction	c. 拆除
4. formwork	d. 模板，模板工程
5. topography	e. 搅拌厂
6. reinforcement	f. 混凝土
7. concrete	g. 钢筋，加固
8. mixing plant	h. 地形，地貌
9. sanitary	i. 污水；下水道；污物
10. layout	j. 卫生的，清洁的

★ Ⅲ. Read the following pictures.

1. Find a proper word or phrase to describe each of the pictures.

1)________________________________

2)________________________________

3)________________________________

4)________________________________

5)______________________________

6)______________________________

2. Read the pictures again and tell your partners which ones are related to each other and what kind of construction work they show.

Part Two

扫一扫，听录音

Site Analysis

Before you begin work on a building site you must be familiar with all the details of the site. You must analyze its location, its limitations and its features.

The following items should all be included in an analysis of the building site. To ensure that you don't forget any of these it may be useful to make your own checklist.

Location

The location of a site should include the suburb, street and lot number for future reference.

Title Search

A title search can be made to establish the correct size, shape and position of an allotment. This can be done by applying to the regional Lands Titles Office. You will require the owner's name and / or the address of the site to obtain the plan number. You may then look at the plan in the office on a micro film, or purchase the plan for a small fee.

Zoning

It is important to determine the zoning of an allotment before planning commences. Zoning will vary from area to area and controls what the land may be used for. You can't, for instance, build a block of flats on an allotment that is zoned R1 which only allows single dwellings to be built.

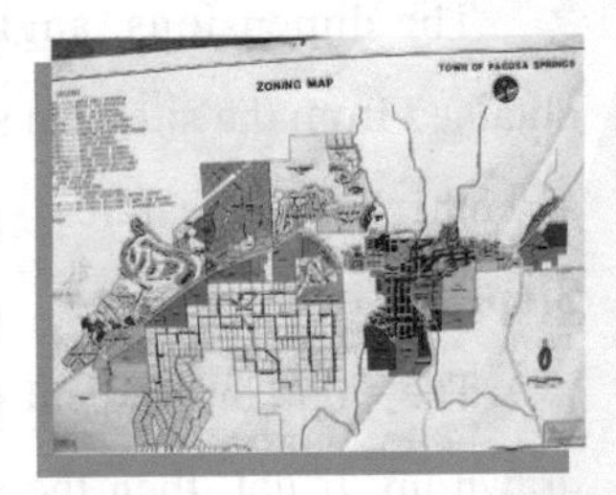

Easements

Easements are areas of land, or part of an allotment, reserved by law for a specific use or purpose, such as access, drainages or municipal services. As part of providing services to an allotment, a service main may need to be placed down the side or across the block of the allotment. This easement will be shown and registered on the Certificate of Title to which it refers. An owner is not permitted, by law, to build upon any land forming part of a registered easement if the building prevents access.

Encumbrance

An encumbrance is a restraining covenant which the developer may have placed on the site to restrict the style, color or profile, etc, of any development. For example, the estate may not permit the use of corrugated iron for any fencing or sheds.

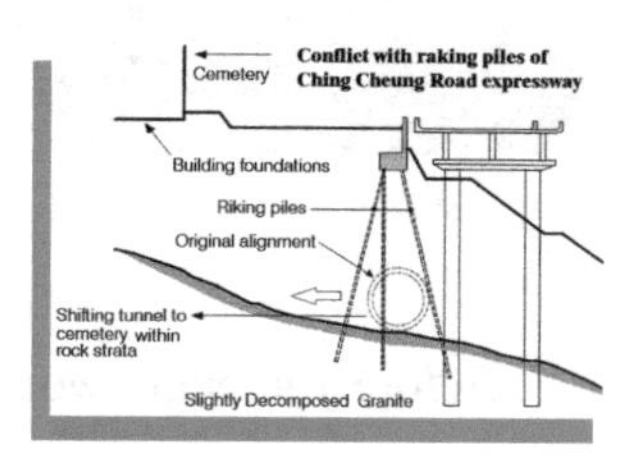

Location of Boundaries

When you build it is important that you are building within the boundaries on the current allotment. The location of the boundaries, apart from the description shown on the Certificate of Title, can be found by reference to the survey pegs. If these can not be located, then a surveyor may be required to accurately locate and peg the site boundaries, including any easements or other site restriction. The surveyor obtains the dimensions, angles and other details of the allotment by reference to the Certificate of Title. The corners of the site are usually identified by white painted pegs.

Dimensions

The dimensions, angles, areas, lot number and orientation of site can all be obtained from the site plan shown on the Diagram from Certificate of Title, along with any other special features.

Slope or Gradient

The slope or gradient of a site will be shown on a contour plan if one has been drawn up. If not, then the surveyor or builder will be required to survey the site to determine the rise and fall of the land. This is important to both the designer and builder because it will determine perhaps the style of house or type of footing used, or may determine the extent of cut and fill required if a conventional type of building is to be constructed. In any case, the slope of the land should be determined at the planning stage.

Access

Access to the site is not only a requirement for the owner but it is also an important issue for the builder. You should think about delivery of materials, particularly concrete or bricks, and the opportunity for any excavation plant to gain access. Provisions for storage of materials on site must also be considered.

Existing Features

The presence of existing buildings, trees, vegetation, fences or any special features must be described in a site investigation. This gives the designer, engineer, or builder every opportunity to plan and cost the building works.

Existing buildings may need to be demolished or built around and therefore may add considerably to the cost of the job. The location, size and type of any trees must be shown on the site plan. In some cases large trees may not be allowed to be removed, or if they are they may significantly affect the soil they were taken from. The removal of the butt and roots may require considerable excavation which will require filling to be properly prepared and compacted.

Vegetation

The removal of vegetation from the site will be necessary to give a clear area for the building. Some properties may have extensive amounts of vegetation to be removed and this must be allowed for in the cost of preparing the site. Fences may be an issue if the proposed building is to be constructed right on the boundary line. If new fences are to be erected, the cost of removing the old ones must be accounted for.

Conspicuous Features

Any other features on the allotment must be shown on the site plan, to allow the greatest opportunity possible to plan the building and its construction and eliminate any potential problems. Features such as a water course, evidence of filling, the existence of any excavation or retaining wall may all have a bearing on the initial design of the construction.

注：easement 地役权（对房产使用权的限制规定）

Professional Words and Expressions

allotment [ə'lɒtmənt]	*n.*	供分配使用的土地
zoning ['zəʊnɪŋ]	*n.*	土地（功能）区划，分区制
commence [kə'mens]	*v.*	开始
dwelling ['dwelɪŋ]	*n.*	住宅，寓所
easement ['i:zmənt]	*n.*	地役权，供役权
municipal [mju:'nɪsɪpl]	*adj.*	市政的，市的
main [meɪn]	*n.*	总管道
encumbrance [ɪn'kʌmbrəns]	*n.*	限定
covenant ['kʌvənənt]	*n.*	契约
estate [ɪ'steɪt]	*n.*	房地产
orientation [ˌɔ:riən'teɪʃn]	*n.*	方位
gradient ['greɪdiənt]	*n.*	梯度，倾斜度
contour ['kɒntʊə(r)]	*n.*	地形，等高线
footing ['fʊtɪŋ]	*n.*	基脚
excavation [ˌekskə'veɪʃn]	*n.*	挖掘
provision [prə'vɪʒn]	*n.*	装置，设备，临时设施
demolish [dɪ'mɒlɪʃ]	*v.*	拆除
compact [kəm'pækt]	*v.*	夯实，压紧，（使）坚实
conspicuous [kən'spɪkjuəs]	*adj.*	明显的
course [kɔ:(r)s]	*n.*	流向
lot number		地块（地段）编号，批号
Lands Titles Office		土地所有权办公室
corrugated iron		瓦楞铁，铁皮波纹瓦
survey peg		测量标桩
water course		水流，水系；水路
retaining wall		挡土墙

Exercises

★ Ⅰ. Reading Comprehension

1. Describe why it is important for a builder to know the slope of a building site.

2. Describe why the access to the site is required.

3. Describe the problems caused by the removal of large trees from the area where a dwelling is to be constructed.

4. Describe how the title search is obtained.

5. Describe why it is important for a builder to make sure of the correct boundary alignment on a building site before work begins.

★ Ⅱ. Translation

A. Translate the following sentences into English.

1. 施工前，必须熟悉施工现场。

2. 现场通道不仅是雇主的要求，对施工者也是十分重要的。

3. 土地区划因地域不同而变化并控制土地的用途。

4. 现场调查必须描述现场的现有建筑、树木、植被、围栏或其他特别的事物。

5. 如果要修新围墙，则必须计算旧围墙的拆除费用。

B. Translate the following paragraph(s) into Chinese.

A site appraisal is one of the first stages in planning the construction of a building. Information gathered from this evaluation is not only needed by the builder but is also required by any architect or engineer working on the project. You must have a good knowledge of the following before commencing work on a site: 1) Location, which may include title, zoning, easement and encumbrance; 2) Description of site, which may include boundaries, dimensions, slop/gradient, access, existing features and vegetation; 3) Conspicuous features as mentioned in Site Analysis; 4) Services available, which you must also consider, such as electricity, water, sewer, gas, telephone, etc.. Which services are already available and what will it cost to connect those? Which are not? 5) Soil investigation, which can be complex and should be undertaken by a soil engineer. It takes into consideration not only the composition of the ground but also many factors which might influence the moisture content of a foundation.

★ Ⅲ. Group Activities

1. Generally, there are three stages in construction: pre-construction stage, construction stage and post-construction stage. Please find out what should be done in each stage.

2. Discuss and list the information you must have before commencing work on site.

扫一扫，听录音

The Pride of Chinese Architecture

The Palace Museum: Established in 1925, the Palace Museum is located in the imperial palace of the consecutive Ming and Qing Dynasties. Situated in the heart of Beijing, the Palace Museum is approached through the Gate of Heavenly Peace (*Tian'anmen*). Ancient China's astronomers endowed the location with cosmic significance. Because of its centrality and restricted access, the palace was called the Forbidden City. With rich collections representing the broad spectrum of 5,000 years of Chinese civilization, the Palace Museum is committed to the preservation of national heritage and looks forward to carrying on the legacy of the past for future generations.

Self-assessment

Now I know

· site investigation can be classified into____________________;

· preliminary site works will concern ____________________;

· site clearance includes ____________________;

· site layout includes ____________________;

· site analysis concerns many items such as ____________________

____.

建筑英语

English for Building Construction

工作手册

班级 ____________
学号 ____________
姓名 ____________

中国建筑工业出版社

Unit One

★ Ⅰ. Learning Objectives

- ◆ To improve listening, reading and writing skills;
- ◆ To learn the professional words and expressions in the unit;
- ◆ To understand drawings and documents used on site;
- ◆ To use the correct terminology in the description of building.

★ Ⅱ. Before Learning

Task 1: You are required to observe a building and write down its structure, building elements and their functions.

____

____

____

____

Task2: Before learning this unit, make sure you understand the following terms or expressions.

☐ tender
☐ client
☐ contractor
☐ consultant
☐ manufacture
☐ subcontractor
☐ drawing
☐ specification
☐ letter of acceptance
☐ performance security

★ Ⅲ. While Learning

Task1: Listen to the conversation and fill in the blanks with what you hear.

1) ____________
2) ____________
3) ____________
4) ____________
5) ____________
6) ____________
7) ____________
8) ____________

Task2: Drawings and documents.

1) Drawings include assembly drawing, component drawing, location drawing, perspective drawing, floor plan, site layout, survey drawing, elevation, section,

structure, etc.

2) Documents include Bill of quantities, daywork sheet, insurance policy, licence, programme, schedule, soil report, specification, etc.

Read the following pictures and write down your understanding of these pictures.

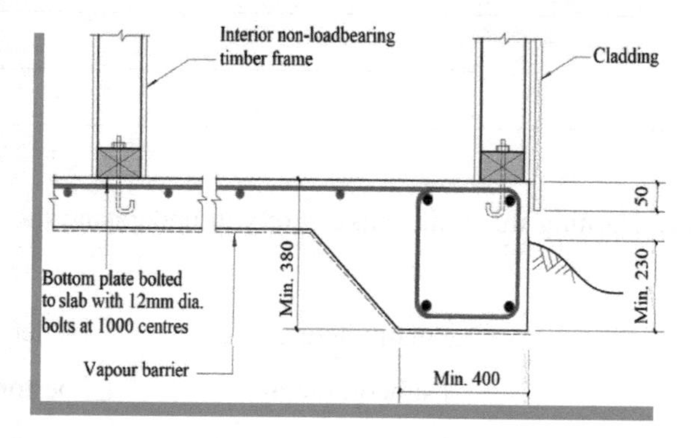

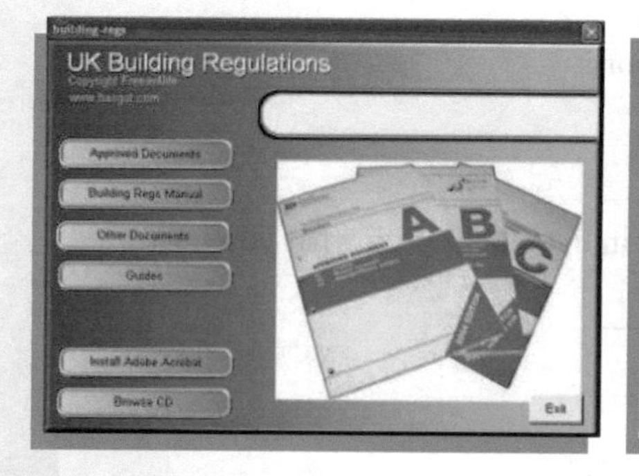

	Task Name	Duration	Priority	Predecessors
1	Demolition	6 d	Lowest	
2	Site Preparation	6 d	Lowest	1
3	Cast-In-Place RC Pile	20 d	Lowest	2
4	Excavation & Support System	30 d	Lowest	3
5	Foundation Baseplate	8 d	Lowest	4
6	RC Formwork	42 d	Lowest	5
7	Steel Formwork	30 d	Lowest	6
8	Roof works	6 d	Lowest	7
9	Water supply & sewerage works	30 d	Lowest	7
10	Power supply system	30 d	Lowest	7
11	Lighting system	20 d	Lowest	7
12	Air conditioning	30 d	Lowest	7
13	Computer & communication network	30 d	Lowest	7
14	Floor finish & polishing	60 d	Lowest	8
15	Internal wall finish	30 d	Lowest	14
16	External wall finish	20 d	Lowest	8
17	Internal partition wall	30 d	Lowest	9,10,11,12,13
18	Ceiling work	40 d	Highest	15
19	Site improvements	6 d	Lowest	18
20	Landscaping work	6 d	Lowest	19

Task3: Building elements are divided into primary and secondary elements.

1) Primary elements include _______, _______, _______, _______, _______.

2) Secondary elements include _______, _______, _______, _______, _______.

3) Read the following picture and list the building elements and their functions.

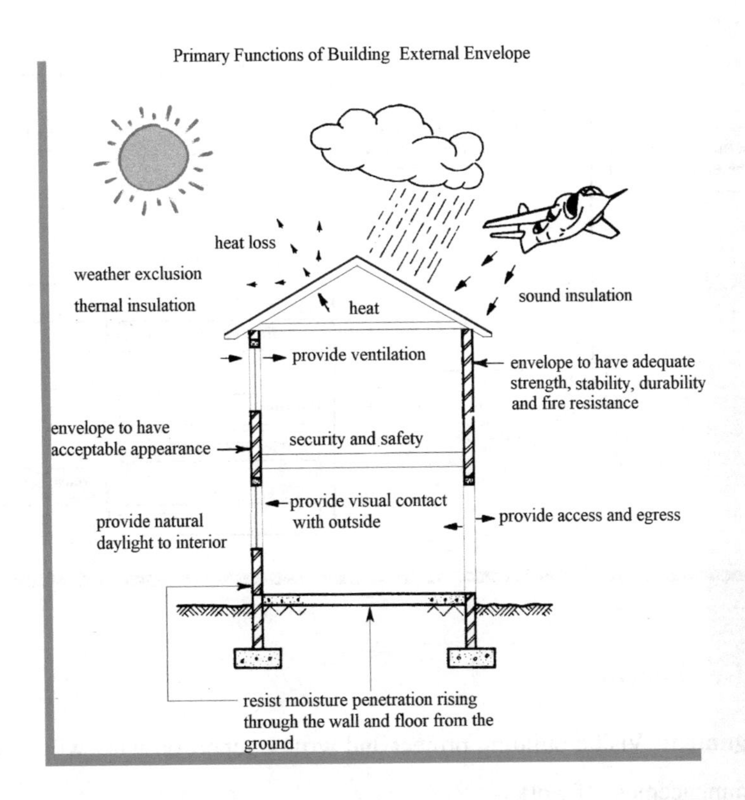

Building Elements	Functions

Task4: Read the unit again and complete the following Mind Map.

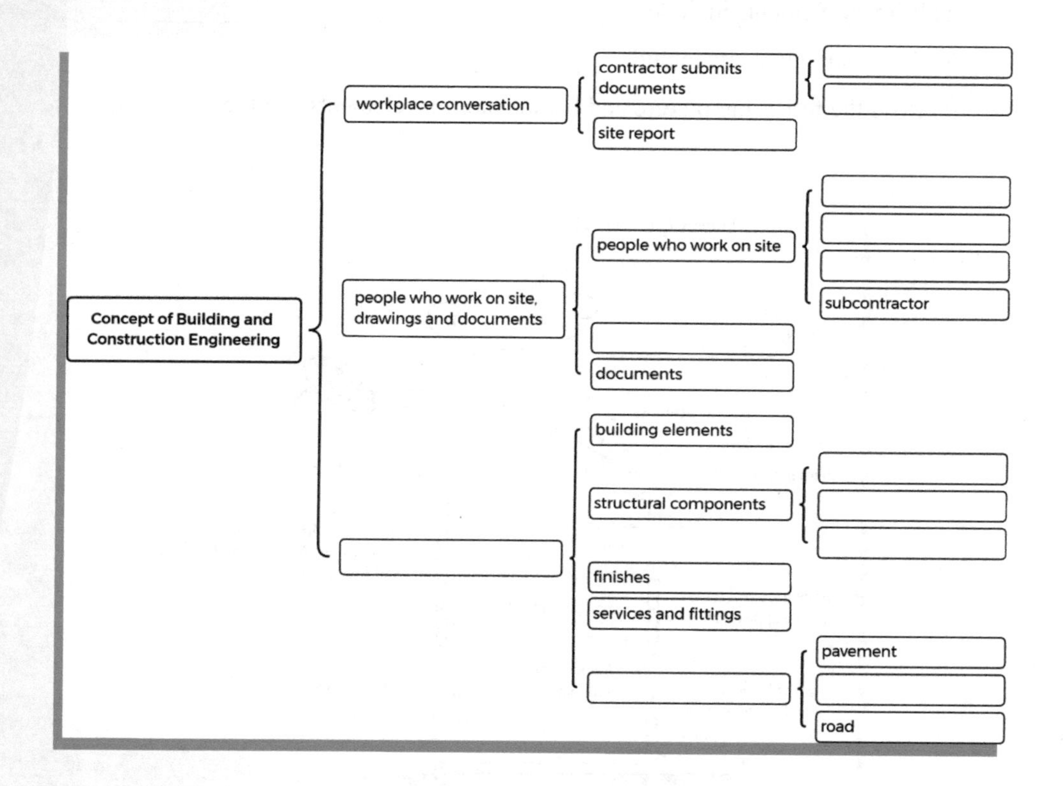

★ Ⅳ. After Learning

Assignment: Visit a building project and write a report on what will be prepared before commencement of works.

____

____

____

____

____

____

____

____

____

____

____

____

Unit Two

★ Ⅰ. Learning Objectives

◆ To improve listening, reading and writing skills;

◆ To learn the professional words and expressions in the unit;

◆ To understand preliminary site work;

◆ To use the correct terminology in the description of preliminary site work and site analysis.

★ Ⅱ. Before Learning

Task 1: You are required to visit a building site and write down what should be considered before the construction work starts.

____

____

____

____

Task2: Before learning this unit, make sure you understand the following terms or expressions.

☐ topography　☐ allotment　☐ contour

☐ layout　☐ easement　☐ land title office

☐ reinforcement　☐ covenant　☐ survey peg

☐ sanitary　☐ excavation

★ Ⅲ. While Learning

Task1: Listen to the conversation and fill in the blanks with what you hear.

1) ____________　5) ____________

2) ____________　6) ____________

3) ____________　7) ____________

4) ____________　8) ____________

Task2: Concept of preliminary site work.

The purpose of the investigation is to achieve better design, economical preliminary work and better construction method. Site investigation will usually be conducted in the two aspects: above ground investigation and below ground investigation.

1) Work in above ground investigation includes:

· ______________________, such as obstruction, existing building and topography;

· ______________________, such as neighboring properties;

· ______________________, e.g. nearest town for facilities & labour.

2) Work in below ground investigation includes:

· ______________________ to determine strength of soil and to detect presence of causes, mines and harmful materials;

· ______________________such as water pipes, sewage pipes, electrical cables, telephone lines, and gas pipes underground;

Task3: Preliminary site works.

Preliminary site works usually involves site clearance and site layout.

The demolition work such as existing structure, trees, and other existing services should be done during the _______________ stage.

The access, storage facilities, workshops or fabrication areas and temporary services should be considered during the_______________ stage.

Task4: Site analysis.

Read the pictures below and write down your understanding for them.

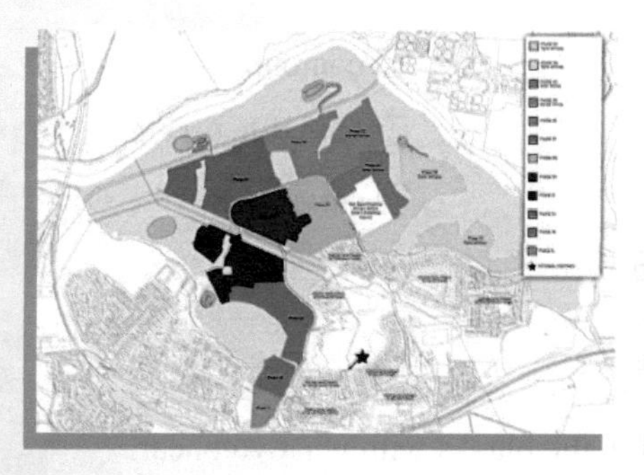

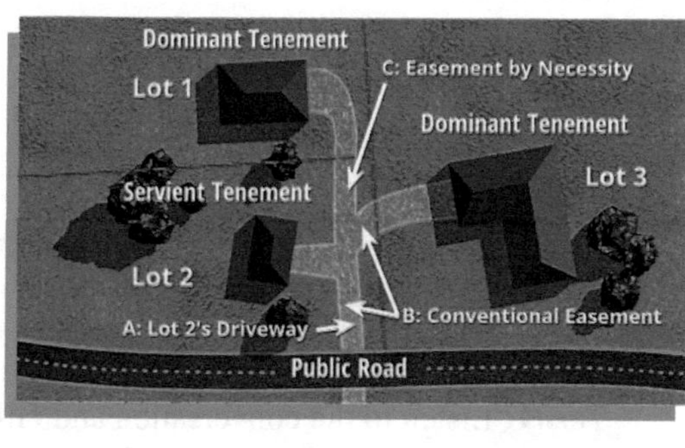

____

____

____

____

____

____

____

Task5: Read the unit again and complete the following Mind Map.

- Preliminary Site Work
  - workplace conversation
    - how to build on a crowded site
    - site report
  - concept of preliminary site work
    - site investigation
      - ______
      - ______
    - preliminary site works
      - ______
      - ______
  - ______
    - location
    - ______
    - encumbrance
    - location of boundaries
    - ______
    - slope or gradient
    - ______
    - existing feature
    - ______
    - conspicuous features

★ Ⅳ. After Learning

Assignment: Your company is taking a construction work in the following site to make it into a resident-commercial block, and your boss assigned you to do the site investigation work. finish the note sheet in which tips are given for the investigation work. Read the samples and finish the rest sheet.

Tips for site investigation

Work to be done	Notes
Existing buildings	3 Resident blocks
Obstruction	none

Unit Three

★ Ⅰ. Learning Objectives

◆ To improve listening, reading and writing skills;

◆ To learn the professional words and expressions in the unit;

◆ To understand foundation types, foundation construction and foundation movement;

◆ To use the correct terminology in the description of substructure.

★ Ⅱ. Before Learning

Task 1: You are required to observe the substructure on the construction site and write down the types of substructure, particularly, the types of foundation and think about the requirements of a satisfactory foundation.

____

____

____

____

Task2: Before learning this unit, make sure you understand the following terms or expressions.

☐ shallow foundation ☐ deep foundation ☐ strip foundation

☐ raft foundation ☐ pile foundation ☐ pad foundation

☐ foundation movement ☐ plastic soil

☐ pre-tensioned ☐ shear force

★ Ⅲ. While Learning

Task1: Listen to the conversation and fill in the blanks with what you hear.

1) ____________ 5) ____________

2) ____________ 6) ____________

3) ____________ 7) ____________

4) ____________ 8) ____________

Task2: Types of foundation and requirements for foundation construction.

1) Foundation can be generally divided into shallow foundation and deep

foundation on terms of its depth, also into strip foundation, raft foundation, pile foundation and pad foundation according to its shape.

2) A satisfactory foundation must be safe against a structural failure that could result in a collapse, feasible both technically and economically, practical to build without adverse effect to surrounding property, and must not settle in such a way as to damage the structure.

Read the following pictures and write down your understanding of these pictures.

____

____

____

____

Task3: Foundation settlement.

1) Three types of foundation settlement are ________, ________, _______.

2) Read the following pictures and analyze their causes.

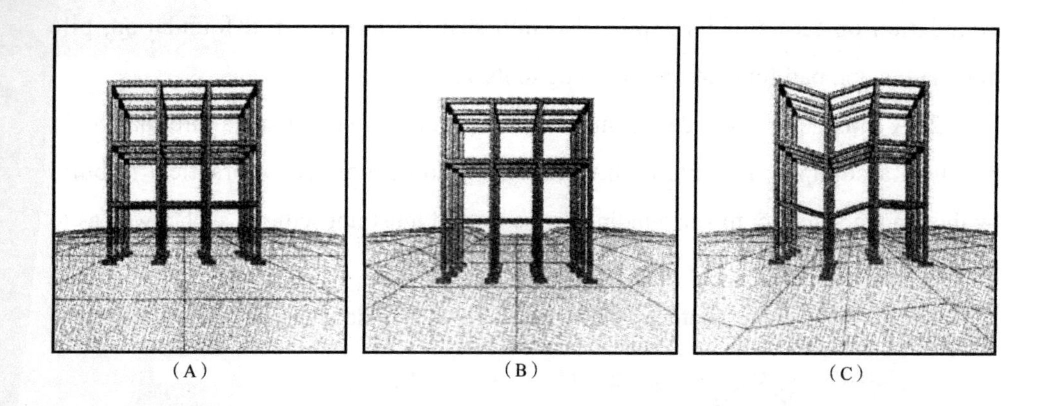

(A) (B) (C)

Types of settlement	Causes

Task4: Read the unit again and complete the following Mind Map.

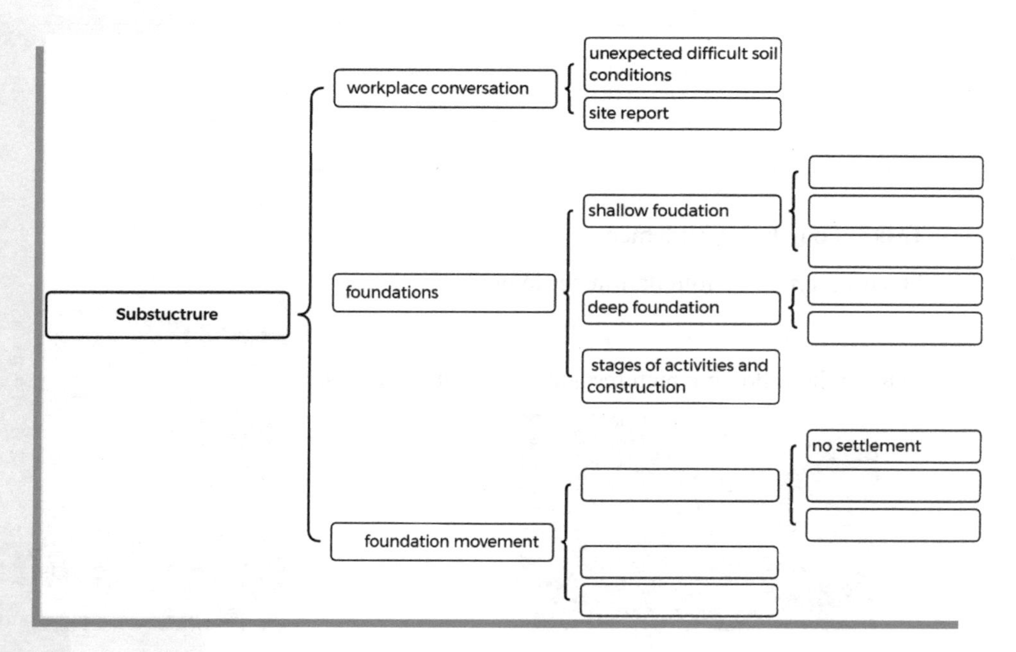

★ Ⅳ. After Learning

Assignment: Visit a building project and write a report on foundation construction.

____

____

____

____

____

____

____

____

____

Unit Four

★ I. Learning Objectives

◆ To improve listening, reading and writing skills;

◆ To know some basic skills in concrete construction;

◆ To learn the professional words and expressions in the unit;

◆ To get knowledge of basic structural principles;

◆ To use the correct terminology in the description of superstructure.

★ II. Before Learning

Task 1: You are required to observe our teaching building and explain the characteristics of its superstructure.

____

____

____

____

Task2: Before learning this unit, make sure you understand the following terms or expressions.

□ superstructure	□ live loads	□ stress
□ framed structure	□ column	□ deflection
□ solid structure	□ frame	□ deformation
□ dead loads	□ force	

★ III. While Learning

Task1: Listen to the conversation and fill in the blanks with what you hear.

1) ____________	5) ____________
2) ____________	6) ____________
3) ____________	7) ____________
4) ____________	8) ____________

Task2: Framed structure.

1) In solid structure, the wall is loadbearing and it takes the loads from the roof and floors and transmit them to the foundation. In framed structure, the loads from

roof and floors transmit to beams and columns, then to foundations, and the walls are enclosing elements.

2) Structural materials for framed structures can be steel, concrete reinforcement cast-in-situ or precast, prestressed concrete, or timber.

Read the following pictures and write down your understanding of these pictures.

____

____

____

____

____

____

Task3: We need to observe the basic structural principles in building construction.

1) The dead load on a structure mainly include: ________, ________, ________, ________, ________.

2) The live load on a structure mainly include: ________, ________, ________, ________, ________.

3) Read the following table and analyze the function or possible consequence of the elements on a structure.

Elements	Function or Consequence
force	
compressive force	
shear force	
stress	
deformation	
strain	
deflection	

Task4: Read the unit again and complete the following Mind Map.

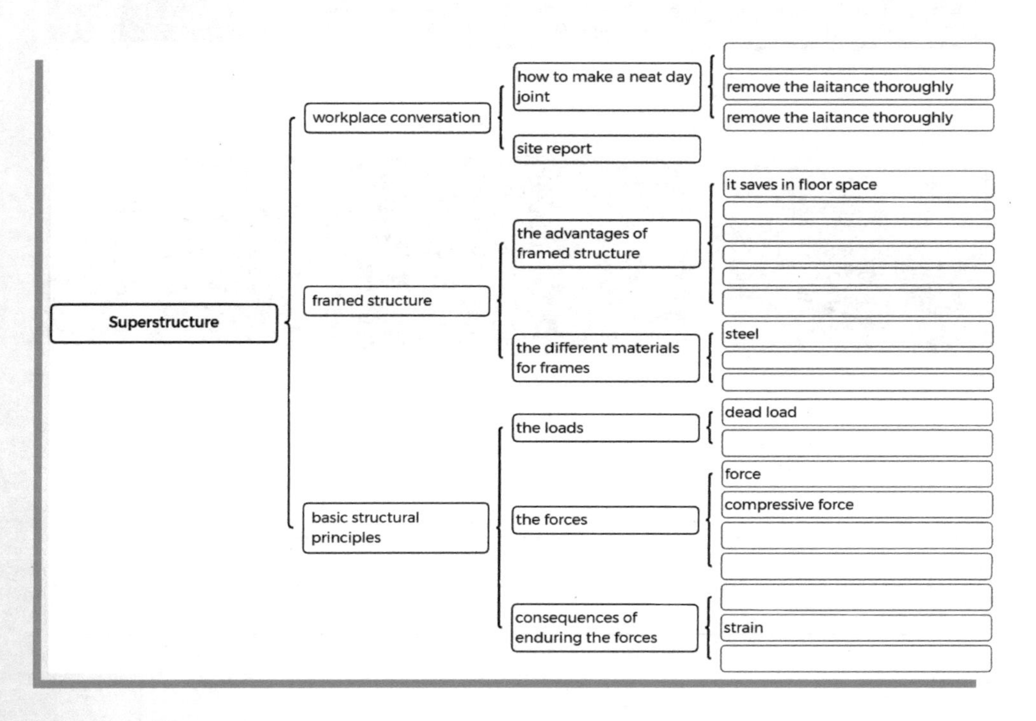

★ Ⅳ. After Learning

Assignment: Observe a building on campus and describe the characteristics of the structure, and analyze the loads it takes and the structural principles it follows.

____

____

____

____

____

Unit Five

★ Ⅰ. Learning Objectives

- ◆ To improve listening, reading and writing skills;
- ◆ To learn the professional words and expressions in the unit;
- ◆ To have basic understanding of roof construction;
- ◆ To use the correct terminology in the description of roof structure.

★ Ⅱ. Before Learning

Task 1: You are required to observe the roof of a building and write down its function and performance requirements.

____

____

____

____

Task 2: Before learning this unit, make sure you understand the following terms or expressions.

□ ridge	□ insulation	□ fire resistance
□ truss	□ drainage	□ air-conditioning plant
□ roo fing	□ ventilation	
□ stability	□ cladding	

★ Ⅲ. While Learning

Task1: Listen to the conversation and fill in the blanks with what you hear.

1) ____________	5) ____________
2) ____________	6) ____________
3) ____________	7) ____________
4) ____________	8) ____________

Task2: Roof construction.

1) The performance requirements for roofs include: strength and stability, weather exclusion, thermal insulation, sound insulation, fire resistance, drainage, durability, lighting, ventilation & smoking dispersion, and appearance.

2) There are many types of roofs of different shapes, pitches, materials and spans. Read the following pictures and write down your understanding of these pictures.

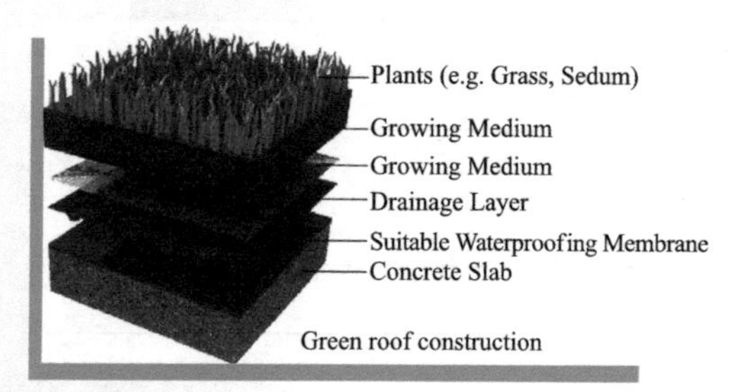

Green roof construction

____

____

____

____

____

____

____

____

Task3: Roof strutting is a very important part of the roof frame.

1) Several factors in fluence the type of roof used, including ______________, ______________________, __________ and ___________.

2) The ________, __________ and ____________ of all roof struts must be constructed to transfer the loads to the foundations in the most direct route.

3) Read the following pictures and list the general construction rules of roof strutting.

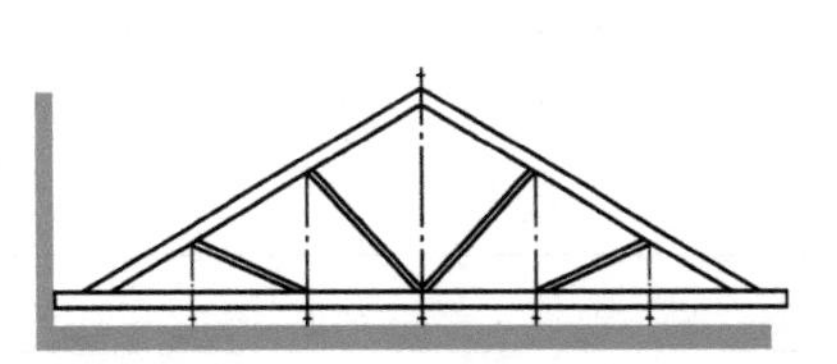

Construction Conditions	Rules of Roof Strutting

Task4: Read the unit again and complete the following Mind Map.

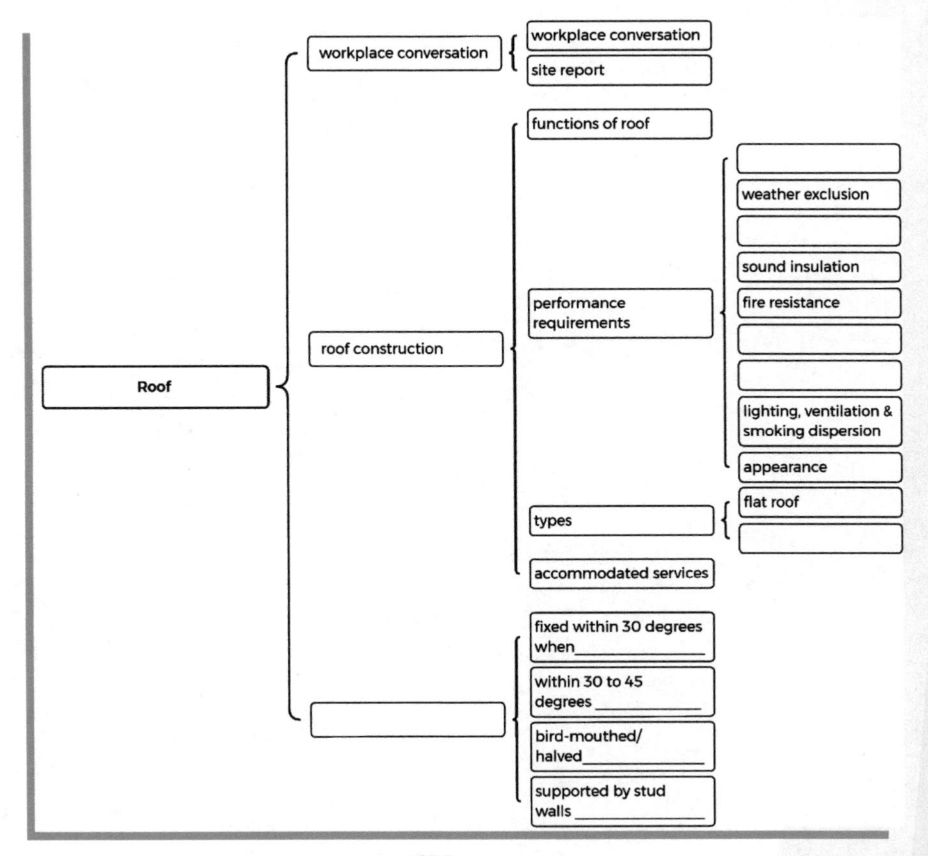

★ Ⅳ. After Learning

Assignment: Visit a building project and write a report on what to be done to meet the performance requirements of roofs.

____

____

____

____

____

____

____

____

Unit Six

★ Ⅰ. Learning Objectives

- To improve listening, reading and writing skills;
- To know some basic skills in laying bricks;
- To learn the professional words and expressions in the unit;
- To get knowledge of wall constructions;
- To use the correct terminology in the description of walls.

★ Ⅱ. Before Learning

Task 1: You are required to observe the walls in a classroom and explain the functions of the walls in different parts of the classroom.

____

____

____

____

Task2: Before learning this unit, make sure you understand the following terms or expressions.

☐ brick work	☐ sound absorber	☐ shrinkage
☐ partition	☐ separating wall	☐ expansion
☐ specification	☐ masonry wall	
☐ sound reduction	☐ monolithic wall	

★ Ⅲ. While Learning

Task1: Listen to the conversation and fill in the blanks with what you hear.

1) ____________ 5) ____________

2) ____________ 6) ____________

3) ____________ 7) ____________

4) ____________ 8) ____________

Task2: Wall constructions.

1) Walls can be classified into load bearing walls, non-load bearing walls,

external non-load bearing walls, partitions, separating walls, compartment walls and retailing walls. Different walls are of different functions and follow different construction requirements.

2) In construction, the walls can be masonry wall built with individual brick/block cemented together, monolithic wall built with concrete, frame wall made of timber of metal, membrane wall composed of a core and two thin skins of plastic, metal or plywood.

Read the following pictures and write down your understanding of these pictures.

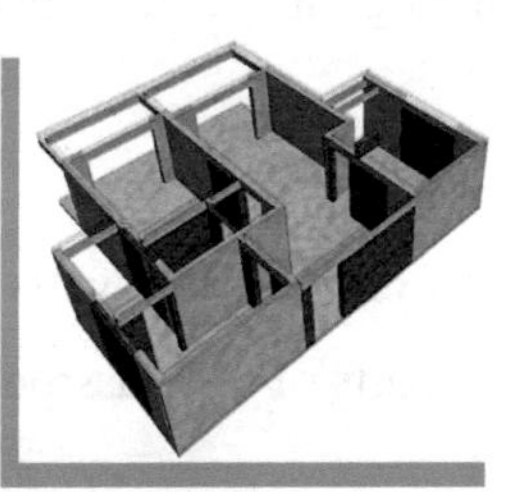

____

____

____

____

____

Task3: The design and construction of load bearing walls.

1) The crucial factors to be considered in designing a load bearing wall include: ____________, ____________, ____________,____________.

2) The factors affect the stability of a masonry wall include: _______, _______, _______, _______, _______.

3) To help the walls withstand the lateral loads, we can take the following measures: ______________, ______________, ______________.

4) To help the walls withstand the vertical loads, we can take the following

measures: ______________, ______________, ______________.

5) In brick wall construction, we should pay attention to the following factors: ______________, ______________, ______________, ______________.

Task4: Read the unit again and complete the following Mind Map.

- Walls
  - workplace conversation
    - the way to lay bricks
      - ______
      - comply with specification
      - ______
    - site report
  - wall constructions
    - the definition of walls
    - the different materials of walls
    - the different types of walls
      - load bearing walls
      - ______
      - ______
      - ______
      - ______
      - retaining walls
    - the functions of walls
      - to enclose the structural frame
      - ______
      - ______
      - ______
      - ______
      - ______
      - ______
    - the constructions of walls
      - masonry wall
      - ______
      - ______
      - ______
  - load bearing walls
    - the factors to be considered in designing load bearing wall
      - the condition of loading
      - ______
      - ______
      - ______
    - the ways to withstand the lateral loads on walls
      - ______
      - ______
      - the mortar joints
    - the ways to resist the vertical loads on walls
      - ______
      - properly bonded cross walls
      - ______
    - bricks walls construction

★ Ⅳ. After Learning

Assignment: Observe a building on campus and describe the functions of walls in different part of the building.

____

____

____

____

____

Unit Seven

★ I. Learning Objectives

- ◆ To improve listening, reading and writing skills;
- ◆ To learn the professional words and expressions in the unit;
- ◆ To understand floor construction;
- ◆ To use the correct terminology in the description of floor construction.

★ II. Before Learning

Task 1: You are required to observe the floors in a teaching building, and write down the answer to the following questions:

1) What are the functions of floors in a structure?

2) How will a floor be constructed?

____

____

____

____

____

Task2: Before learning this unit, make sure you understand the following terms or expressions.

☐ bitumen	☐ asphalt	☐ verandah
☐ topping	☐ pouring	☐ porch
☐ plank	☐ fill	☐ raft slab
☐ camber	☐ moisture	☐ floating slab

★ III. While Learning

Task1: Listen to the conversation and fill in the blanks with what you hear.

1) ____________ 5) ____________

2) ____________ 6) ____________

3) ____________ 7) ____________

4) ____________ 8) ____________

Task2: Concept of floors.

Fill in the blanks with the correct answer from the passage.

1) Floors are ________________ structural elements.

2) The types of floor construction are determined by two factors: __________ and ___________.

3) In the construction of floor ______________ is the main material, and ________________are widely used.

4) The functions of floors include _______________ and _________________.

5) The floor at ground and basement is called _________________, while the upper floor which spans between two supports is called _______________.

Task3: Floor construction.

Read the list of numbers and find out what they refer to in the studying material.

150 mm	
100-150 mm	
25-50 mm	
25-65 mm	
50-75 mm	
50 mm	

Task4: Site analysis.

Read the pictures and fill in the blanks.

Picture No.1 ____

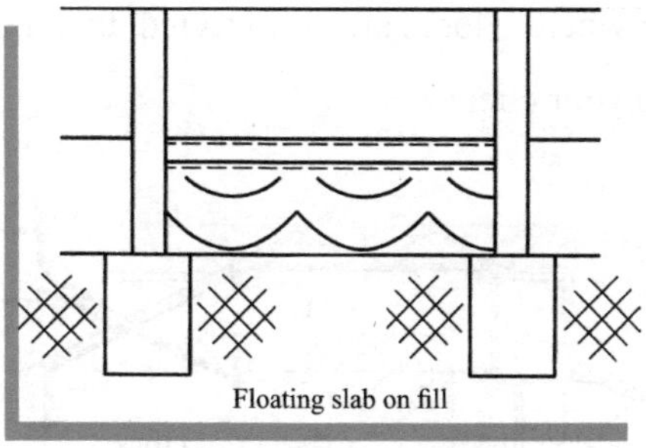

Floating slab on fill

Picture No.2: ____

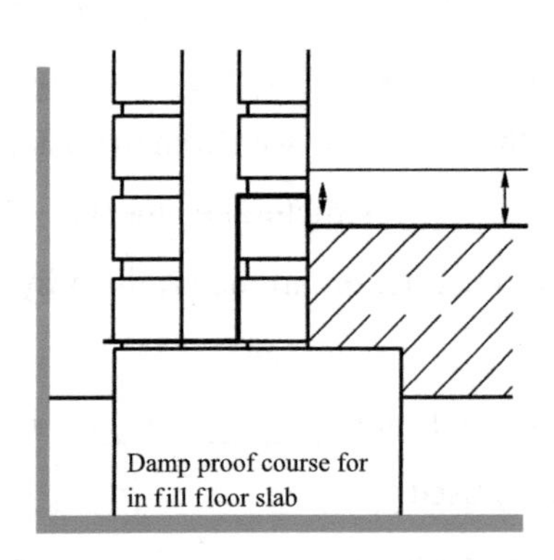

Task5: Read the unit again and complete the following Mind Map.

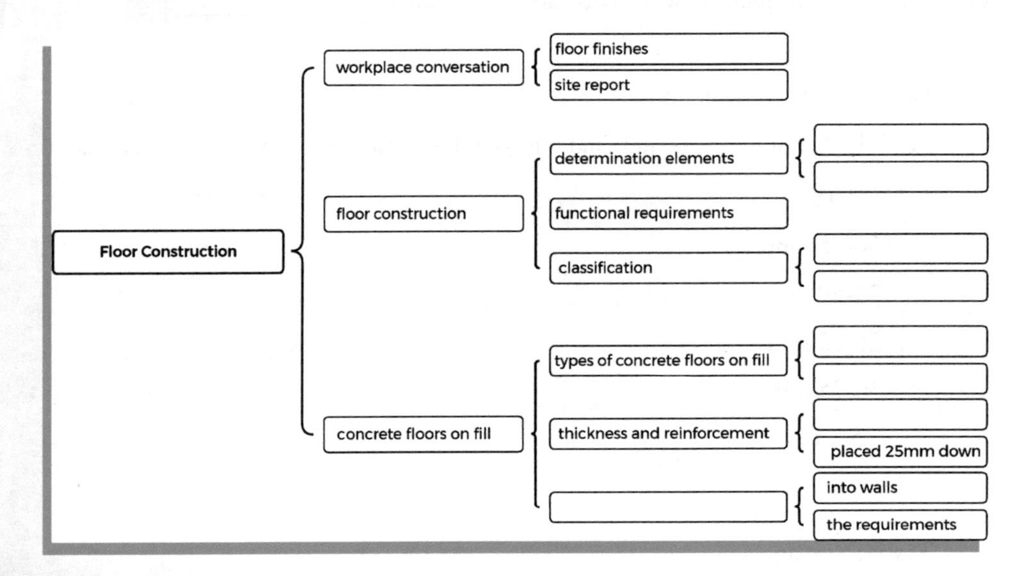

★ Ⅳ. After Learning

Assignment: Your company is going to bid for the construction of a two-storey house. It is your responsibility to explain to the client about the floor systems. First mark on the section plan where floors are constructed, then make some notes for your incoming presentation to your client.

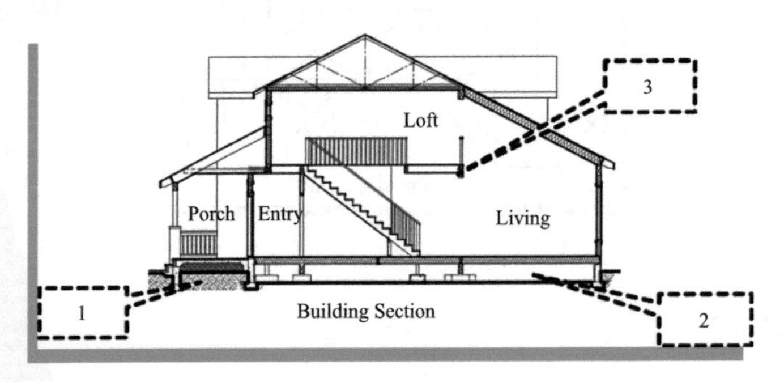

Notes:

Type of Floor	Tips for Presentation
1.	
2.	
3.	

Unit Eight

★ Ⅰ. Learning Objectives

◆ To improve listening, reading and writing skills;

◆ To learn the professional words and expressions in the unit;

◆ To understand demountable partitions & suspended ceilings, paint and painting system;

◆ To use the correct terminology in the description of finishes.

★ Ⅱ. Before Learning

Task 1: You are required to observe the ceiling, partition and coating of a building, and write down their functions.

____

____

____

____

____

Task2: Before learning this unit, make sure you understand the following terms or expressions.

☐ demountable partition ☐ module frame ☐ soffit
☐ strip ☐ step ladder ☐ ceiling grid
☐ suspension hangers ☐ ducting
☐ format ☐ baffle

★ Ⅲ. While Learning

Task1: Listen to the conversation and fill in the blanks with what you hear.

1) ____________ 5) ____________
2) ____________ 6) ____________
3) ____________ 7) ____________
4) ____________ 8) ____________

Task2: Demountable partition and suspended ceilings.

1) The majority of demountable partition systems consist of a framework and infill panels.

2) Suspended ceiling systems offer even more advantages for building construction, including a range of acoustical control options, fire protection, esthetic appearance, flexibility in lighting and HVAC (heating, ventilation, and air conditioning) delivery, budget control and optional use of overhead space.

Read the following pictures and write down your understanding of these pictures.

____

____

____

____

____

Task3: Finishes and coatings.

The type of paint and the appropriate painting system depends on:

1) ____.

2) ____.

3) ____.

4) ____.

5) Read Part Two and find out the corresponding information.

Key Terminology	Information
Sealers	
Primers	
Priming coat	
Undercoats	
Finishing coats	

Task4: Read the unit again and complete the following Mind Map.

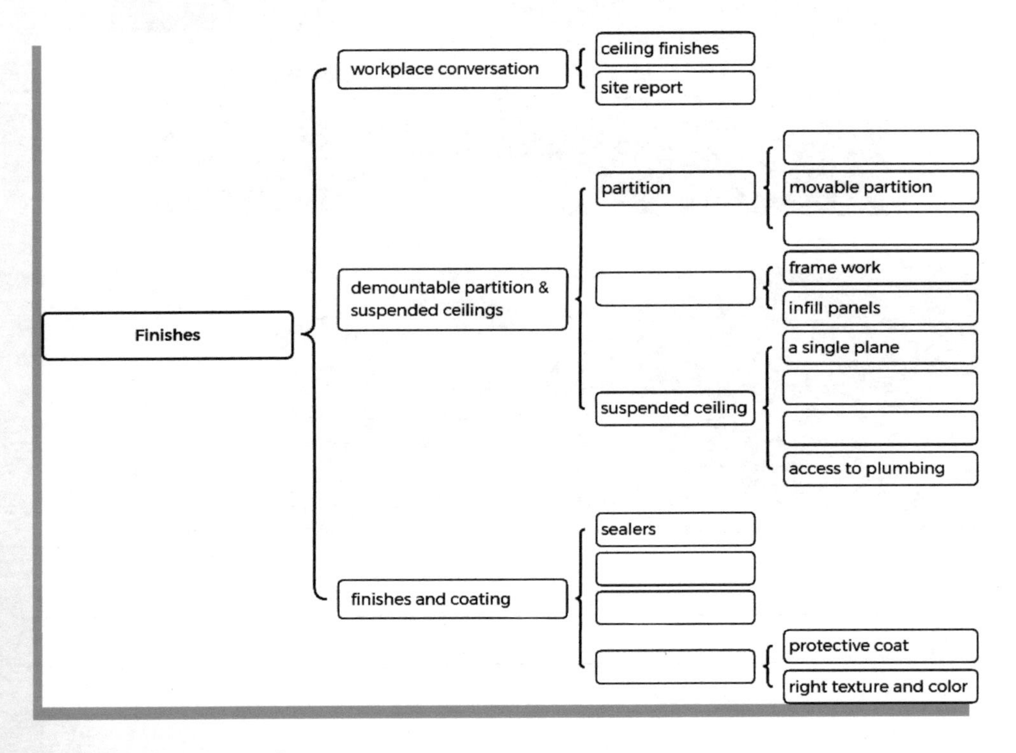

★ Ⅳ. After Learning

Assignment: Visit a building project and write a report on the roles and functions of different materials in a decoration system.

Unit Nine

★ Ⅰ. Learning Objectives

- ◆ To improve listening, reading and writing skills;
- ◆ To learn the professional words and expressions in the unit;
- ◆ To understand the importance of electrical and plumbing installation;
- ◆ To use the correct terminology in the description of building services.

★ Ⅱ. Before Learning

Task1: You are required to observe a building and write down its services and their functions.

____

____

____

____

____

Task2: Before learning this unit, make sure you understand the following terms or expressions.

□ ductwork	□ electrician	□ constant flow rate
□ honeycomb	□ air conditioning	□ controller valve
□ intumescent material	□ duct-mounted silencer	□ plumber
□ flexible ducting	□ fire damper	

★ Ⅲ. While Learning

Task1: Listen to the conversation and fill in the blanks with what you hear.

1) ____________	5) ____________
2) ____________	6) ____________
3) ____________	7) ____________
4) ____________	8) ____________

Task2: Building services include environmental services, utility services and building services systems.

1) Environmental services include _________, _________, _________.

2) Utility services include __________, __________, __________, __________, __________, __________, __________, __________.

3) Building services systems include__________, __________, __________.

Read the following pictures and write down your understanding of these pictures.

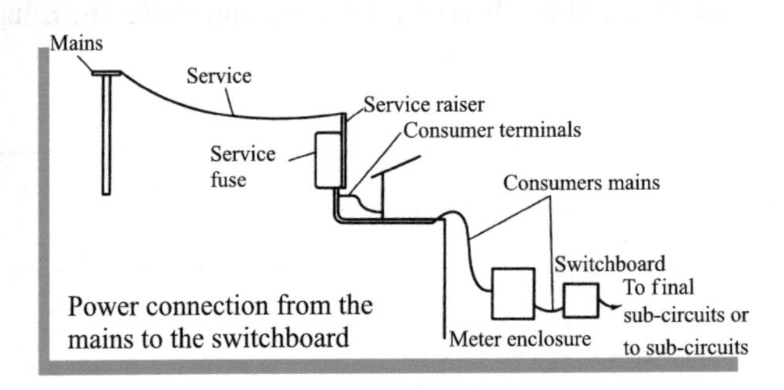

Power connection from the mains to the switchboard

____

____

____

____

____

Task3: Section B Part Two-plumbing installation.

1) A plumber must be registered to install the sewerage system, hot and cold reticulated water systems and gas services.

2) Read the passage and find the corresponding answers of plumbing installation.

Key Terminologies	Corresponding Answers
Sewerage plumbing and drainage	
Cold water reticulation	
Mains pressure units	
Reduced pressure units	

Task4: Read the unit again and complete the following Mind Map.

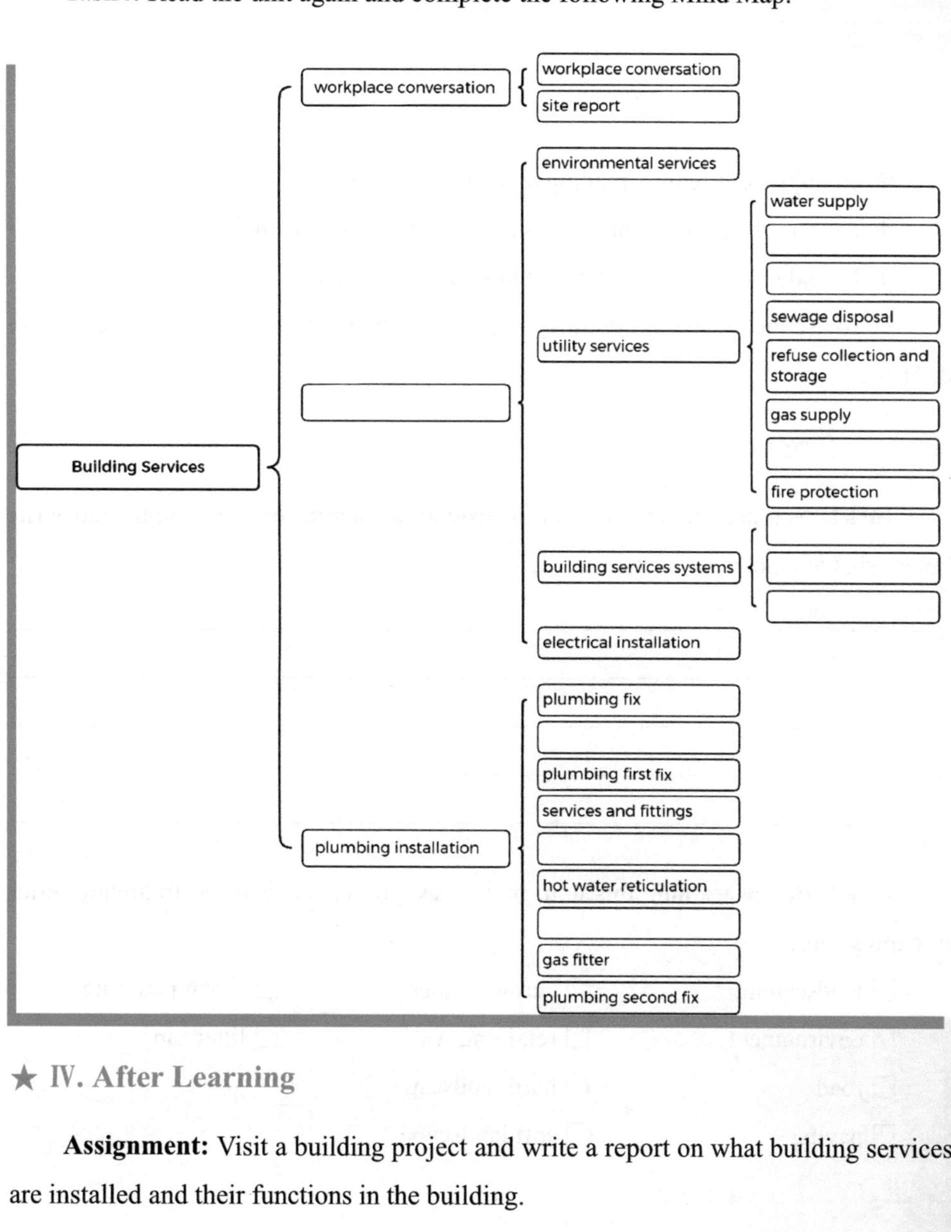

★ Ⅳ. After Learning

Assignment: Visit a building project and write a report on what building services are installed and their functions in the building.

____

____

____

____

____

____

____

____

Unit Ten

★ Ⅰ. Learning Objectives

◆ To improve listening, reading and writing skills;

◆ To learn the professional words and expressions in the unit;

◆ To understand external works and surface drainage;

◆ To use the correct terminology in the description of external works and surface drainage.

★ Ⅱ. Before Learning

Task1: You are required to visit a residential quarter or the campus and write down what are included in external works.

____

____

____

____

____

Task2: Before learning this unit, make sure you understand the following terms or expressions.

☐ landscaping　☐ surface water　☐ urban planning

☐ environment　☐ retaining wall　☐ litter bin

☐ road　☐ hard landscape

☐ paving　☐ soft landscape

★ Ⅲ. While Learning

Task1: Listen to the conversation and fill in the blanks with what you hear.

1) ____________　5) ____________

2) ____________　6) ____________

3) ____________　7) ____________

4) ____________　8) ____________

Task2: External works.

External works will generally include the following:

· paving to driveways and paths,

· disposal of surface water,

· retaining walls where the site is cut-and- fill,

· landscaping gardening.

Read the following pictures and write down your understanding of these pictures.

____

____

____

____

____

Task3: Paving materials are important for external works.

1) The main considerations when choosing paving materials are:

______________________________, ______________________________,

______________________________, ______________________________,

______________________________, ______________________________.

2) F ive common types of paving materials include:

__________, __________, __________, __________, __________.

3) Read the following pictures and name the paving materials and their functions.

Paving Materials	Functions

Task4: The slab heights above paving level may vary depending on:

- ________________________,
- ________________________,
- ________________________,
- ________________________,
- ________________________.

Task5: Read the unit again and complete the following Mind Map.

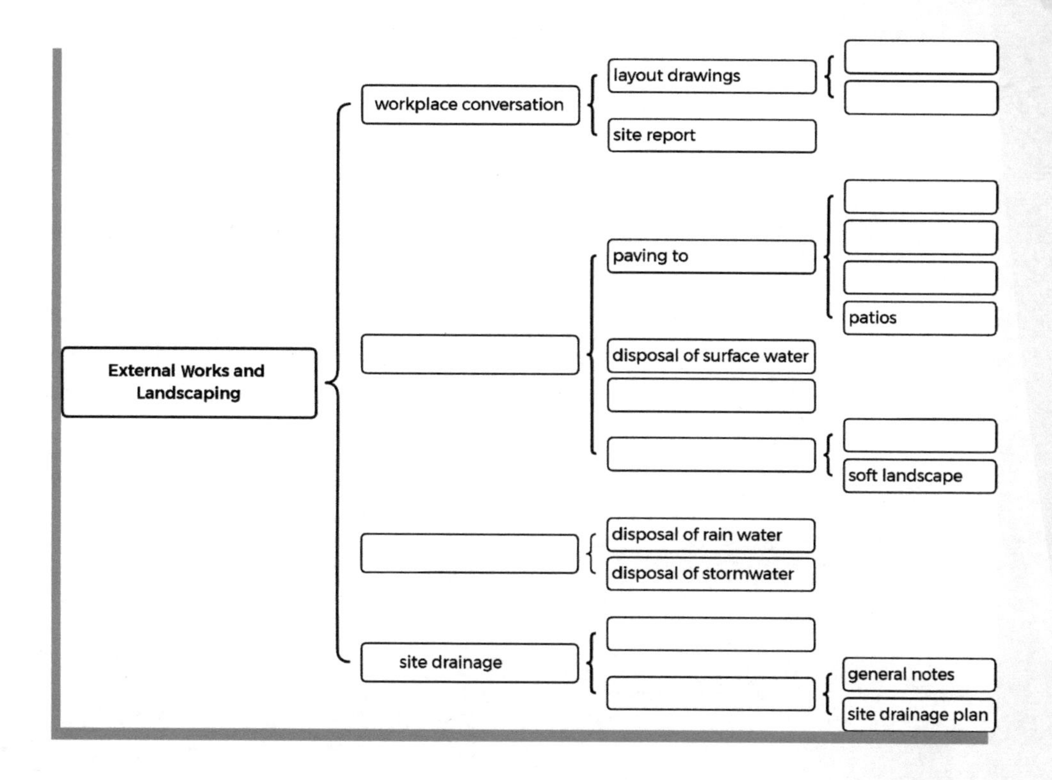

★ IV. After Learning

Assignment: Visit a residential quarter or consult a landscape architect and write a report on what will be involved in external works.

____

____

____

____

____

____

____

Now I can

☐ use the correct terms in preliminary work;

☐ write a site report according to the workplace conversation;

☐ describe what to be done in site investigation and preliminary site works;

☐ distinguish the items in site analysis;

☐ conduct a proper site planning.

Unit Three

Substructure

Learning Objectives

After learning this unit, you will be able to

- know foundation types;
- write a site report according to the workplace conversation;
- understand foundation construction and activities;
- get knowledge of foundation movement;
- use the correct terminology in the description of foundation;
- understand Chinese excellent architectural culture.

扫一扫，听录音

Part One

Section A Workplace Conversation

Professional Words and Expressions

unearth [ʌn'ɜ:θ]	*n.*	发掘，掘出
sand [sænd]	*n.*	砂
excavate ['ekskəveɪt]	*v.*	挖掘，开挖
trench [trentʃ]	*n.*	渠，壕，沟（槽，渠）
backfill ['bækfɪl]	*v.*	回填
leanmix ['li: nmɪsk]	*n.*	贫拌合料，少灰混合
pour [pɔ:(r)]	*v.*	灌，浇筑（混凝土）
outcrop ['aʊtkrɒp]	*n.*	砂层，层
strip footing		条形基础

Excavation (unearthing a surprise)

If excavations for foundations unexpectedly show difficult soil conditions, the structural engineer has to decide quickly what should be done.

★ I. Listen to the conversation and fill in the blanks with what you hear.

(Tony—Structural Engineer; Peter—Site Agent)

Tony: Hmm, yes, I see the problem. We didn't expect to find 1)____________ here, did we? Well, we'll certainly have to do something about it. There are supposed to be strip footings here, I think, aren't there?

Peter: Yes, that's right.

Tony: Well, I'm afraid it won't be much good if we just leave it. That sand could easily 2)____________________ under the foundations.

Peter: Yes, that's just what I thought. There'll be seven storeys going here, you know, so it'll be 3)____________________.

Tony: Yes. And you say you've 4)____________ the trench to the usual depth of one meter? So how far down did you come across this sand, then?

Peter: Oh, it must have been about 5)_________________ down when we first saw it.

Tony: I see. Well, first I think you ought to excavate another 400 millimeters to see whether you can 6)____________________.

Peter: Right you are.

Tony: If you do manage to, then you can backfill with leanmix concrete before you pour your foundations.

Peter: Okay. And 7)______________________________? Do you think we should have a wider footings or something?

Tony: Yes. If the outcrop is still there, then obviously I'll have to redesign that footing. You'll have to make it wider, 8)______________________________.I think I'd better have a drawing ready for you tomorrow in case it's needed.

★ Ⅱ. Peter writes a site report according to the conversation above.

This morning we started excavating and found a problem. ____________________

____

____

____

____

____

____

扫一扫，听录音

Section B　Foundations

Shallow Foundations of a House Versus Deep Foundations of a Skyscraper

The foundation is that part of the building which is in direct contact with the ground and which transmits the load of the building to the underlying soil. Foundations are generally divided into two categories: shallow foundations and deep foundations.

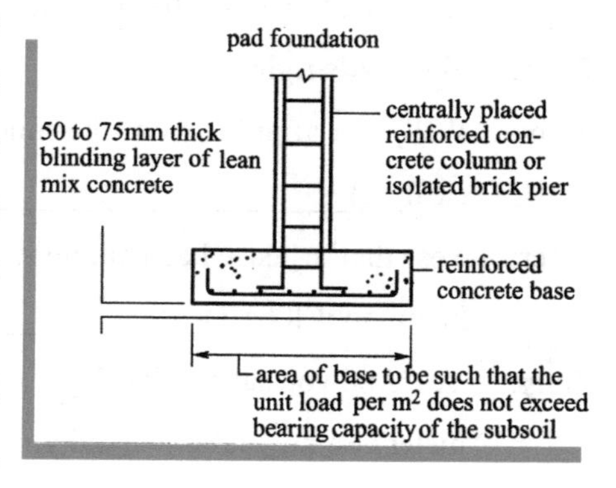

Shallow foundations, often called footings, are usually embedded about a meter or so into soil. One common type is the spread footing which consists of strips or pads of concrete (or other materials) which extend below the frost line and transfer the weight from walls and columns to the soil or bedrock.

Another common type of shallow foundation is the slab-on-grade foundation (or raft foundation) where the weight of the building is transferred to the soil through a concrete slab placed at the surface. Slab-on-grade foundations can be reinforced mat slabs, which range from 25 cm to several meters thick, depending on the size of the building, or post-tensioned slabs, which are typically at least 20 cm for houses, and thicker for heavier structures.

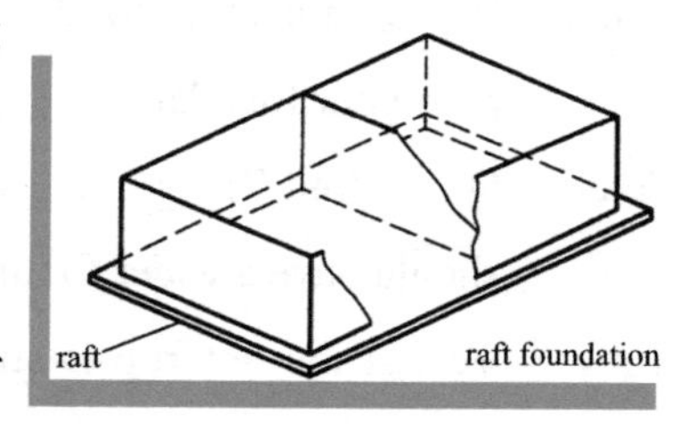

A deep foundation is used to transfer the load of a structure down through the upper weak layer of topsoil to the stronger layer of subsoil below. There are different types of deep footings including impact driven piles, drilled shafts, caissons, helical piles and earth stabilized columns. The naming conventions for different types of footings vary between different engineers. Historically, piles were wood, later steel, reinforced concrete, and pre-tensioned concrete.

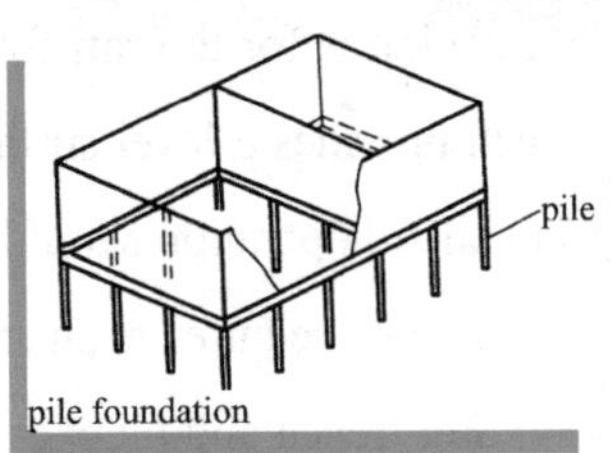

A satisfactory foundation for a building must meet three general requirements.

1. The foundation, including the underlying soil and rock, must be safe against structural failure that could result in collapse.

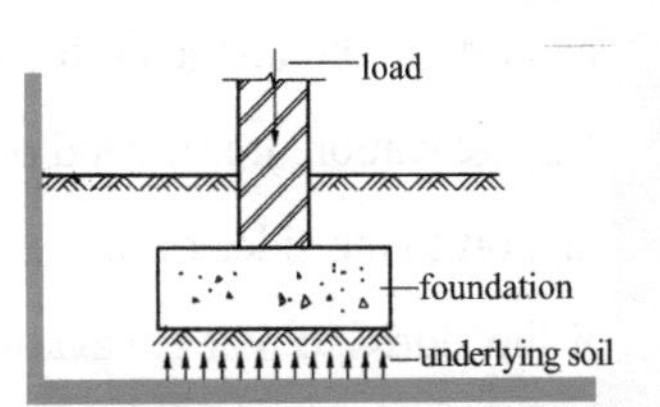

2. During the life of the building, the foundation must not settle in such way as to damage the structure or impair its function.

3. The foundation must be feasible both technically and economically, and practical to build without adverse effect to surrounding property.

Therefore, the foundation engineer is responsible for assessing the factors that affect the choice of a foundation type for a building, working together with other members of the design and construction team to select the most suitable foundation system.

Stages of Activities and Construction Involved in Foundation:

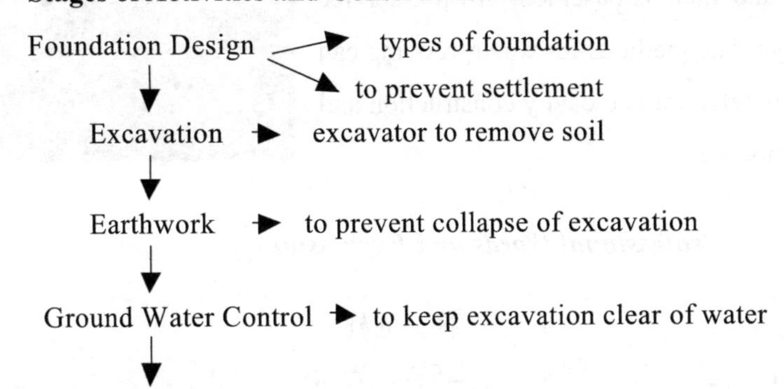

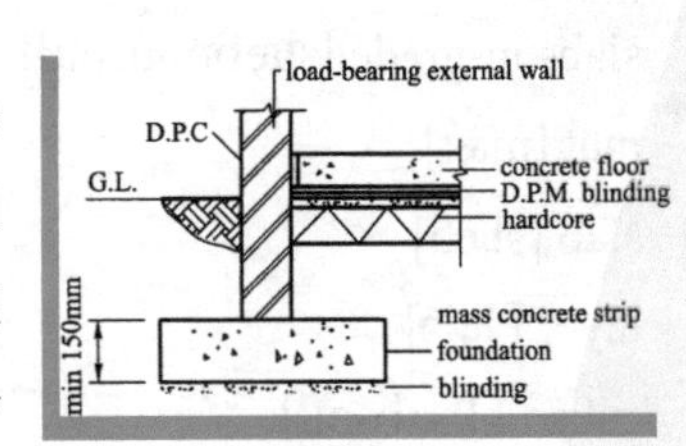

The basic design considerations are that the foundation must be of sufficient area to spread the total load as not to overstress the soil. From the calculated total loads of the building and known bearing capacity of the subsoil, the size of the foundation can be easily determined.

Excavation is removal of soil in the ground for foundation and basements. Types of excavation depend on the types of foundations. Trench excavation may be used for the construction of strip foundation and laying of pipes and cables. Pit excavation is used for pad foundation, pile caps, stanchions, lamp posts, lift shafts, etc. Reduced level excavation is for the raft foundation. The excavation is to reduce the level of the ground which provides a level surface from which construction may take place. For sloping site, cut and fill operation should be carried out to achieve a level surface. Therefore, basement construction requires open excavation.

For trench and pit excavation, the vertical face without support can be adequate for good strong soils. However, in this excavation, it is a good practice to provide some form of support to the sides of the excavation to prevent the collapse of the sides, so that excavation is safe for the excavators to work in.

Ground water control is important during excavation to enable workmen engaged in substructure work to carry out their duties in suitable conditions and to enable the bottoms of excavations to be properly prepared to the precise levels and dimensions required for the works. The most widely used method of ground water control is pumping from sumps which are temporarily excavated.

Waterproofing is to prevent entry of water into buildings underground such as basement, lift pit, service tunnels and subways. The methods for waterproofing can be of watertight concrete, drained cavity construction and waterproofing membrane.

Professional Words and Expressions

column ['kɔləm]	*n.*	柱，立柱
bedrock ['bedrɒk]	*n.*	基岩，岩床
slab-on-grade [slæbɒngreɪd]	*n.*	地面（混凝土）板，底板式基础
mat [mæt]	*n.*	混凝土垫层
slab [slæb]	*n.*	板
layer ['leiə]	*n.*	层，夹层
subsoil ['sʌbsɔɪl]	*n.*	下层土，老土

caisson ['keɪsən]	*n.*	沉箱，沉井
collapse [kə'læps]	*n.*	崩塌，塌陷
waterproofing ['wɔ:təpru:fɪŋ]	*n.*	防水（层），防水作业
basement ['beɪsmənt]	*n.*	地下室，基础结构
stanchion ['stæntʃən]	*n.*	柱子，标柱，标桩
membrane ['membreɪn]	*n.*	膜（片），隔板，防渗护面，表层
helical ['helɪkl]	*adj.*	螺旋形的，螺纹的
post-tensioned slab		后张法预应力板
shaft [ʃɑ:ft]	*n.*	（竖，升降，通风）井
stabilized [s'teɪbəlaɪzd]	*adj.*	稳定的
pit [pɪt]	*n.*	坑，取土坑，料坑
vertical ['vɜ:tɪkl]	*adj.*	垂直的
sump [sʌmp]	*n.*	集水井，排水沟
drain [dreɪn]	*v.*	排水
cavity ['kævətɪ]	*n.*	（空）腔
earth work		土方工程，土石方工程
frost line		冰冻线，冻深线
shallow foundation		浅基础
deep foundation		深基础
reinforced mat slab		钢筋混凝土筏板或平板
spread footing		大放脚，扩展基础
pre-tensioned concrete		先张法（预应力）混凝土
structural failure		结构损坏，结构性破坏，结构失效
bearing capacity		承重能力，承载力
pile cap		桩承台，桩帽
lift shaft		电梯井，（升降）机竖井
lift pit		电梯井底坑，电梯基坑
service tunnel		地下管道
drained cavity construction		疏水片导流排水法

Notes

1. **The frost line,** also known as frost depth or freezing depth—is most commonly the depth to which the groundwater in soil is expected to freeze. The frost depth depends on the climatic conditions of an area, the heat transfer properties of the soil and adjacent materials, and on nearby heat sources.

2. **Pre-tensioned concrete** is cast around steel tendons—cables or bars—while

they are under tension. The concrete bonds to the tendons as it cures, and when the tension is released it is transferred to the concrete as compression by static friction. Tension subsequently imposed on the concrete is transferred directly to the tendons.

Exercises

★ Ⅰ. Answer the following questions.

1. What is the function of the foundation of a building?
2. What types of foundations are shallow foundations?
3. What types of foundations are deep foundations?
4. What factors should be considered when designing foundations?
5. What are the stages of activities and construction involved in foundation?

★ Ⅱ. Match the following words or phrases with the correct Chinese.

1. basement	a. 柱，立柱
2. stanchion	b. 基岩，岩床
3. waterproofing	c. 斜坡（混凝土）板，踏步式楼板
4. pre-tensioned concrete	d. 工作隧道，辅助隧道，副隧道
5. service tunnel	e. 承重能力，承载量
6. bedrock	f. 先张法（预应力）混凝土
7. slab-on-grade	g. 防水（层），防水作业
8. watertight	h. 底座，基层，基础结构
9. column	i. 柱子，标柱，标桩
10. bearing capacity	j. 不漏水的，防渗的

★ Ⅲ. Read the following pictures.

1. Find a proper word or phrase to describe each of the pictures.

1) 

2) ____________________

3) ______________________ 4) ______________________

5) ______________________ 6) ______________________

2. Tell your class what kind of foundations each picture above shows. Give your reasons in detail.

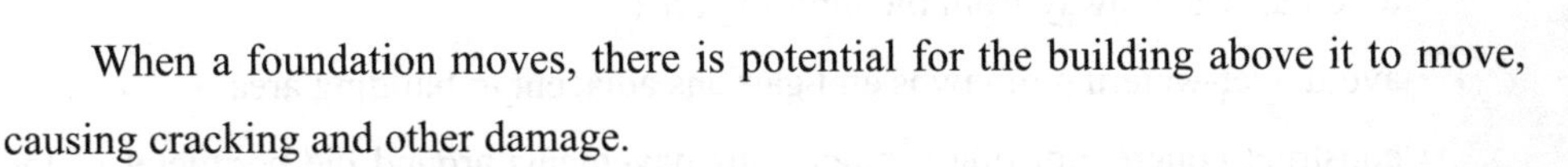

Part Two

扫一扫，听录音

Foundation Movement

When a foundation moves, there is potential for the building above it to move, causing cracking and other damage.

The three main types of foundation movement are:

· settlement

· swelling and shrinking

· shear failure

Settlement

When a building is constructed, its load causes the foundation under it to consolidate. This downward movement is called settlement. All foundations settle to some extent as the soil around and beneath them adjusts itself to the load of the building. If settlement occurs at roughly the same rate from one side of the building to the other, it is called uniform settlement, and no harm is likely to be done to the building. If large amounts of differential settlement occur, the frame of the building may become distorted, floors may slop, walls and glass may crack, and doors and windows may refuse to work properly. Accordingly a primary objective in foundation design is to minimize differential settlement by loading the soil in such a way that equal settlement occurs under the various parts of the building.

In plastic soil, settlement is due to the squeezing of moisture from between soil particles and proceeds slowly. In soils such as sand, settlement is due to the movement of sand particles as they re-align themselves to the load, and takes place comparatively quickly.

Swelling and Shrinking

The heaving pressure that is generated in reactive soils can be great and it is not practical to resist this force. However it is possible to greatly reduce the amount of vertical and lateral movement by keeping the soil at an even moisture content.

When the heaving pressure is uniform over the site, few structural problems are encountered. It is when the movement varies significantly over the site that damaging pressures arise. The differential movement is caused by variations in moisture content, types and depth of soil.

The following practice will assist in reducing building failure in reactive soils:

· divert all water away from the building area

· avoid over-watering of lawns and gardens adjacent to building area

· construct concrete or other impervious pavements around the perimeter of the building.

Shear Failure

Shear failure is likely to occur in both cohesive and noncohesive soils. Noncohesive soils are not compressible, and when overloaded, the soil is forced out from under the footing to bulge upward alongside the footing without moving it. As the soil shrinks on subsequent drying out, the footing drops.

Therefore, foundations are designed to have an adequate load capacity, safely transmit to the soil all the loads from the building so that settlement is even and limited.

(A)Building Before Settlement

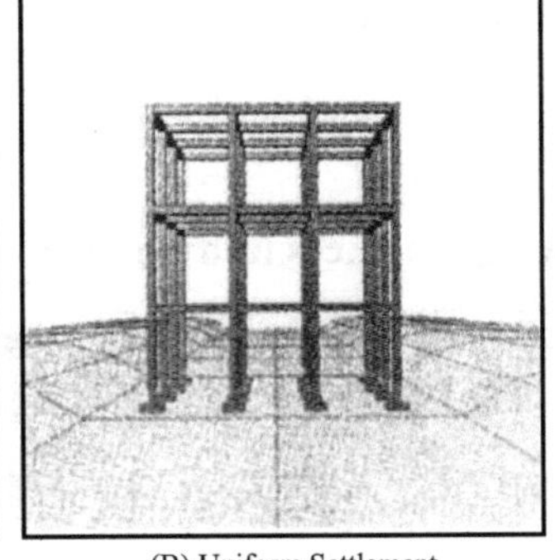
(B) Uniform Settlement

(C) Differential Settlement

Professional Words and Expressions

cracking ['krækɪŋ]	*n.*	裂缝，裂纹
settlement ['setlmənt]	*n.*	沉降，建筑物的下沉
swelling ['swelɪŋ]	*n.*	湿胀，膨胀，隆起
shrinking [ʃrɪŋkɪŋ]	*n.*	收缩
consolidate [kən'sɒlɪdeɪt]	*v.*	固结，加固
uniform ['ju:nɪfɔ:m]	*adj.*	均匀的，匀质的
bulge [bʌldʒ]	*v.*	凸出；膨胀
heaving ['hi:vɪŋ]	*n.*	抬起，隆起
reactive [ri'æktɪv]	*adj.*	活性的
perimeter [pə'rɪmɪtə(r)]	*n.*	周边，边界
shear failure		剪切断裂，剪切破坏
differential settlement		不均匀沉降
plastic soil		塑性土壤，可塑土
lateral movement		侧向变形，侧向移动
moisture content		含水量，含水率，湿度

Exercises

★ I. Reading Comprehension

1. What are the three main types of foundation movement?
2. What is settlement in building construction?
3. How can the differential settlement be minimized in foundation designs?
4. What can usually be done to reduce building failure in reactive soils?

5. How is shear failure likely to occur?

★ Ⅱ. Translation

A. Translate the following sentences into English.

1. 如果建筑物一侧与另一侧的沉降速率大致相同，这叫作均匀沉降，对建筑物可能不会造成损害。

2. 如果发生大的不均匀沉降，会造成建筑物框架扭曲，楼板倾斜，墙体及玻璃破例，门窗不好开关。

3. 基础设计的首要目标就是把不均匀沉降最小化。

4. 通过保持土壤含水量的均衡，就有可能大幅减少地基垂直沉降和侧向沉降。

5. 条形基础或独立基础要构筑到冰冻层下面，将墙柱的重量分布到土壤或基石。

B. Translate the following paragraph(s) into Chinese.

All foundations settle to some extent as the soil around and beneath them adjusts itself to the load of the building. If settlement occurs at roughly the same rate from one side of the building to the other, it is called uniform settlement, and no harm is likely to be done to the building. If large amounts of differential settlement occur, the frame of the building may become distorted, floors may slop, walls and glass may crack, and doors and windows may refuse to work properly. Accordingly a primary objective in foundation design is to minimize differential settlement by loading the soil in such a way that equal settlement occurs under the various parts of the building.

★ Ⅲ. Group Activities

1. How can differential settlement occur? What effects does differential settlement have?

2. What should be considered to limit the building settlement in foundation designs?

扫一扫，听录音

The Pride of Chinese Architecture

National Centre for the Performing Arts (NCPA) is a state-level art institution based in Beijing. It covers an area of 118,900 square metres, with a total area of structure of 217,500 square metres. Its main building is a unique shell-shaped structure. The surface of the building is composed of over 18,000 titanium panels and more than 1,200 pieces of ultra-clear glass, creating a visual effect as a curtain rises on stage.

As China's top performing arts centre, the NCPA will adhere to the guiding principle of "for the people, for art and for the world" and strive for the objectives of being: a key member of prestigious international theatres, the supreme palace of performing arts in China, the leader of arts education and popularization, the grandest platform for international arts exchange, and an important base for cultural and creative industry.

Self-assessment

Now I know

· foundation refers to ______________________________

______________________________;

· foundation is generally divided into ______________________________

______________________________;

· three general requirements for foundation are ______________________________

______________________________;

· stages of activities and construction involved in foundation are ______________________________

______________________________;

· three main types of movement are ______________________________

______________________________.

Now I can

□ use the correct terminology in the description of foundation;

□ write a site report according to the workplace conversation;

□ distinguish shallow foundation and deep foundation;

□ explain the cause for movement.

Unit Four

Superstructure

Learning Objectives

After learning this unit, you will be able to

- know some basic skills in concrete construction;
- write a site report according to the workplace conversation;
- understand framed structure construction;
- get knowledge of basic structural principles;
- use the correct terminology in the description of superstructure;
- understand Chinese excellent architectural culture.

Part One

扫一扫，听录音

Section A Worksite Conversation

Professional Words and Phrases

joint [dʒɔɪnt]	*n.*	接缝，接合处
chippie ['tʃɪpɪ]	*n.*	木工，木匠
batten ['bætn]	*n.*	板条；压条
groove [gru:v]	*n.*	沟，槽
laitance ['leɪtəns]	*n.*	翻沫；水泥浆
shuttering ['ʃʌtərɪŋ]	*n.*	模板
concrete plant		混凝土厂
day joint		施工缝，后接缝
releasing agent		防粘剂，脱模剂

A Crisis on Site

Concrete must be handled carefully to be sure that it hardens at the right time in the right place. A breakdown in supply can cause an awkward problem.

★ I. Listen to the conversation and fill in the blanks with what you hear.

(Tony—Structural Engineer; Malcolm—Concrete Engineer; Mick—Groundworker)

Malcolm: Ah, hello, Tony—I hope you wouldn't be too long. We're in real trouble here.

Tony: Yes, so I was told. What's the matter?

Malcolm: Well, you see. We've poured about half of this slab, and the 1)____________ just broken down. 2) ________________.

Tony: Oh dear. All right, let's see how far you've got. Hmm-well, it's a good job you've been pouring from one side towards the other, isn't it? It's so hard to make 3) ____________ if you've been pouring all over the whole area. Right, stop the pour along the line here, then. I think we'll be able to make 4) ____________.

Malcolm: Okay. We'll get a chippie right away.

Tony: Fine. Now, then, this is what I'd like you to do. Er— first of all, get a batten fixed underneath. That makes a groove, you see, which hides the joint 5) ______________.

Malcolm: Right, I'm with you.

Tony: Okay. Then shutter up the joint, in the form of a step. It helped to make a stronger joint if you do it like that.

Malcolm: I think we should be able to manage that all right.

Tony: Fine. And one last thing, Malcolm, don't leave any 6) ____________ on the concrete, will you? Do be sure you remove it all thoroughly. Get it off while the concrete's still fresh. Otherwise you'll get a weak finish on that concrete, see.

Malcolm: Yes, I know. We'll see to that. I'll stay on late with Mick and do it.

Tony: Oh, and I nearly forgot to say, you will have to remember to apply the 7) ____________ properly. You know, thinly and evenly. Then you'll be able to 8) ____________ tomorrow.

Malcolm: Right, okay. We'll get Mick to do a special job on it, then.

Tony: Yes-well, we'll see how good a job you've made it all tomorrow.

Malcolm: Right, then. Well, thanks very much.

Tony: Oh, that's all right.

★ Ⅱ. Malcolm writes a site report according to the conversation above.

This morning, when pouring concrete, we've got in real trouble. ____________

____

____

____

____

____

Section B Framed Structures

扫一扫，听录音

The structural concepts have been introduced here, namely that of solid structure and skeletal structure(framed structure). In solid structures, the enclosing element which is the wall is loadbearing. The wall takes the loads from the roof and floors and transmits the loads to the foundations (usually strip foundation).

The skeletal structure, requires mainly the wall to be enclosing. Hence, the walls can be quite thin and light in weight and often quite quickly erected. This has become significant for high-rise structures and resulted in the developments of external cladding systems.

Framed structure consists of a skeleton or framework. The loads from roof and floors transmit to beams and columns, then to foundation. This type of structure requires enclosing element (walls) to keep out weather. The advantages of the framed structure are as follows.

- It saves in floor space as walls are thinner especially at lower floors.
- It is flexible in plan and building operation.
- It reduces in dead weight as walls are thin and light.
- It is more economical when roof span and floor span are large.
- It is suitable for both low-rise and high-rise buildings while solid structure is not suitable for 15 storeys or more.
- It is faster in construction.
- It has wider choice of wall materials which can

be masonry, precast concrete slab, metal and glass, etc.

· It makes large openings in walls possible.

The tasks of design and analysis will have to be done many times until a design has been found for the structure that is strong, stable and lasting, to carry safely all loads imposed without deforming excessively under loads. It has to be in suitable forms and use appropriate materials to resist the loads. Stiffening elements are required for the rigidity in frame with wind bracing for high-rise buildings. The design of the structure will also consider fire resistance. The frame has to be maintained long enough for occupants to escape in case of fire. Concrete framed structure is highly fire-resistant while steel framed structure requires fire protection. As for the timber framed structure, the suggestion is maximum three storeys.

Closely spaced columns with short span beams are cheaper than widely spaced columns.

Where possible, the layout of a skeleton should be based on a regular structural grid. The advantages of doing so are as follows.

· Loads are evenly transmitted to foundation, which minimizes relative settlement and makes foundation sizes standardized.

· Beams depths and columns size are standardized, and position of column and beams are the same, which makes framework and size of walls standardized.

· Minimize the use of different reinforcement sizes and greater re-use of formwork, which reduce cost.

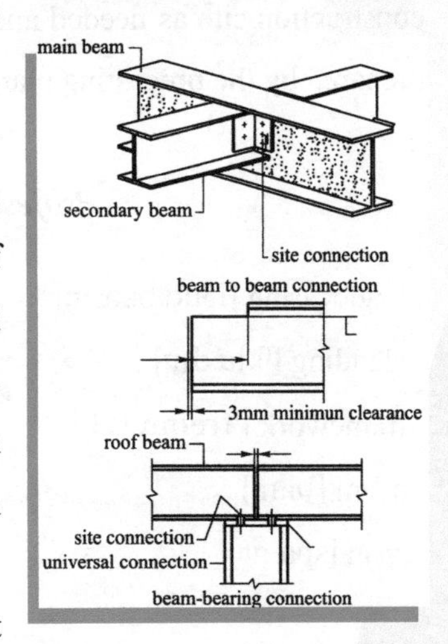

Structural materials can be steel, concrete-reinforcement in-situ or precast, prestrssed concrete, timber-up to three storeys. Steel frame consists of horizontal beams in both directions and vertical columns called stanchions, and standard rolled sections. They are joined together by welding and/or bolts with cleats, and commonly encased in concrete. Beams can carry in-situ or precast floors.

Reinforced concrete frame requires reinforcement

to take tensile stress and formwork to support the concrete until it has gained sufficient strength. It can be cast to any shape and size. Concrete is good in compression but will crack and shear under tension. Therefore, concrete member is designed with main steel reinforcement in areas of tension. Reinforcement must be surrounded by concrete to protect bars against corrosion, fire and to allow the bars to develop its correct force. Reinforced concrete frame has a lot of advantages such as higher fire resistance, regular maintenance not required, greater flexibility in layouts designs, monolithic structure, etc.

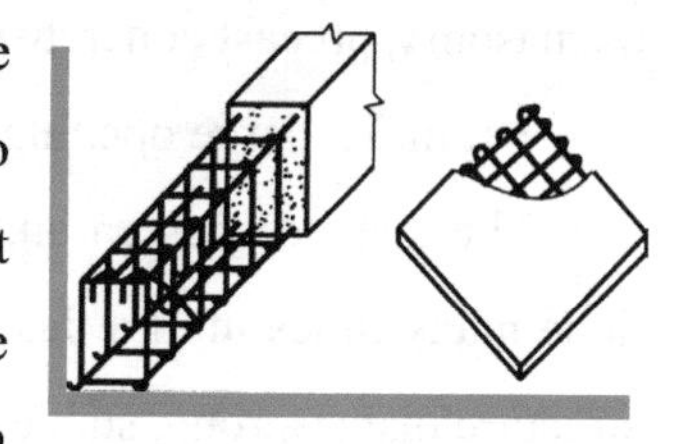

Precast concrete frame developed as a result of attempts to link advantages of steel frame with the economy of the concrete frame. The components cast in one place in formworks, transported and erected into final position in another place. Precast concrete frames can be fabricated in the following ways. The structure drawings for the building are sent to the precasting plant, where engineers and drafters prepare shop drawings that show all the dimensions and details of the individual elements and how they are to be connected. These drawings are reviewed by the engineer and architect for conformance with their design intentions and corrected as necessary. Then the production of the precast components proceeds, beginning with construction of any special molds that are required and fabrication of reinforcing cages, the continuing through cycles of casting, curing, and stockpiling. The finished elements, marked to designate their positions in the building, are transported to the construction site as needed and placed by crane in accordance with erection drawings prepared by the precasting plant.

Professional Words and Expressions

loadbearing [ləud'bεərɪŋ]	*n.*	承载，承重
cladding ['klædɪŋ]	*n.*	外墙板，外墙覆面板；围护结构
framework ['freɪmwɜ:k]	*n.*	框架；构架
beam [bi:m]	*n.*	梁，横梁，承重梁
span [spæn]	*n.*	跨度

masonry ['meɪsənrɪ]	*n.*	砖石；砌体
stiffening ['stɪfnɪŋ]	*n.*	加强，加劲
rigidity [rɪ'dʒɪdətɪ]	*n.*	刚性；刚度
occupants ['ɒkju:pənts]	*n.*	居住者，占有人
in-situ [ˌɪn'saɪtju:]	*adj.*	现场的；原位的
precast [ˌpri:'kɑ:st]	*adj.*	预浇筑的，预制的
prestressed [p'ri:strest]	*adj.*	预应力的
welding [weldɪŋ]	*n.*	焊接法，定位焊接
bolt [bəʊlt]	*n.*	螺栓，螺钉
cleat [kli:t]	*n.*	托座；连接板件
encase [en'keɪs]	*v.*	围住，包住
compression [kəm'preʃn]	*n.*	抗压；压迫
crack [kræk]	*v.*	断裂，折断
shear [ʃɪə (r)]	*v.*	切变；切断
tension ['tenʃn]	*n.*	张力，拉力
corrosion [kə'rəʊʒn]	*n.*	腐蚀，侵蚀，锈蚀
monolithic [ˌmɒnə'lɪθɪk]	*adj.*	整体的；庞大的
cast [kɑ:st]	*v.*	浇筑；铸造
fabricate ['fæbrɪkeɪt]	*v.*	制造；组合
drafter ['drɑ:ftə (r)]	*n.*	草图设计员；绘图员
conformance [kən'fɔ:məns]	*n.*	一致性；符合性
mold [məʊld]	*n.*	模子；模具
curing ['kjʊərɪŋ]	*n.*	固化；养护
stockpiling ['stɒkpaɪlɪŋ]	*n.*	贮存；堆放
designate ['dezɪgneɪt]	*v.*	指明，指出
crane [kreɪn]	*n.*	吊车，起重机
strip foundation		条形基础
precast concrete		预制混凝土
wind bracing		抗风支撑；防风拉筋
rolled section		辗压断面
tensile stress		拉应力
shop drawing		制造图，施工图；深化设计图

Exercises

★ Ⅰ. Answer the following questions.

1. What have been introduced in the passage?

2. What are the differences between solid structure and framed structure?

3. What are the advantages of the framed structure?

4. What does the steel frame consist of and how are their elements joined together?

5. How are the precast concrete frames fabricated and erected?

★ Ⅱ. Match the following words or phrases with the correct Chinese.

1. superstructure	a. 骨架；轮廓
2. harden	b. 吊车，起重机
3. laitance	c. 整体的
4. skeleton	d. 外部的
5. external	e. 翻沫
6. precast concrete	f. 铸造
7. tensile stress	g. 上部结构；上层建筑
8. monolithic	h. 硬化；变硬
9. casting	i. 拉应力
10.crane	j. 预制混凝土

★ Ⅲ. Read the following pictures.

1. Find a proper word or phrase to describe each of the pictures.

1) ______________________________

2) ______________________________

3) ______________________

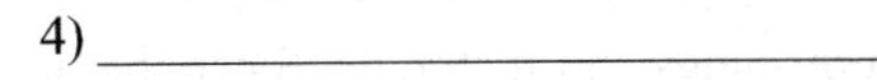

4) ______________________

5) ______________________ 6) ______________________

2. Read the pictures again and tell your partners which ones are related to each other and what kind of construction work they show.

Part Two

扫一扫，听录音

Basic Structural Principles

There are basic structural principles which must be observed in building construction. Will the house stand up to the loads, forces, stresses and strains which are forced upon it by gravity, weather and its own weight? If builders did not observe these structural principles, we could have houses falling down around our ears, causing serious damage to people and property.

Loads

A building's loads can be divided into two types:

- dead load
- live (imposed) load

The dead load on a structure is made up of the weight of the structure including any permanently attached elements that are supported by the walls, such as any floors

or roofs. The dead load also includes the plumbing, heating equipment and those kitchen appliances that would make the house uninhabitable if removed.

The live load (or imposed load) on a structure includes the weight of anything else that may be present in the structure at any time. This includes people, furnishings and materials. Since the imposed load impossible to estimate, it can help up to use a table based on the type of occupancy of a building. Such a table aims to estimate the "probable worst" loading on a structure.

Force

A force is that which produces, or tends to produce, motion or a change of motion of a body. Force is measured in newtons (N) and is equal to 9.81 × mass (in kilograms). For example, a block of concrete with a mass of 2000 kg creates a force of 9.81 × 2000 = 19620 N.

Compressive Force

Compressive force is when a force acts on a body in a manner that tends to shorten the body or to push the part of the body together. The force is a compressive force, and the body (building element) is in a state of compression.

Tensile Force

A tensile force is one which acts on a body in a manner that tends to lengthen the body or to pull the parts of the body apart. The force is a tensile force because the body is in a state of tension.

Shear Force

When two parallel forces having opposite directions act on a body tending to cause one part of the body to slide past an adjacent part, the forces are shear forces.

Stress

A stress is the internal resistance of a body to an external force acting upon it. It is measured by force per area:

$$\text{Stress} = \frac{\text{force}}{\text{Area}}$$

In this equation, force is expressed in Newton (N) and area in square metres (m^2) or square millimeters (mm^2).

Cracking in buildings may occur when the stress placed on a building material

exceeds the strength of that material. This stress may be caused by externally applied loads or internal movement.

Deformation

When a body is subjected to a force, there is change in the shape or size of the body. These changes are called deformations.

When a tensile force is applied to a steel bar, the original length of the rod is increased, and this lengthening is its deformation. This deformation is called strain.

Strain

The deformation of a body under load is called strain. Tensile forces create tensile strain (stretching, elongating or extension). Compressive forces create compressive strain (shortening or contraction).

Deflection

A loaded beam resting on two supports near its ends tends to become concave on its upper surface. We say the beam bends, and because it is downward bending it is called positive bending. The deformation that accompanies the bending is called deflection.

Professional Words and Phrases

stress [stres]	*n.*	应力
strain [streɪn]	*n.*	应变；张力
gravity ['grævətɪ]	*n.*	地心引力；重力
plumbing ['plʌmɪŋ]	*n.*	管道；给排水系统；管件安装
heating ['hi:tɪŋ]	*n.*	供暖；加热
appliance [ə'plaɪəns]	*n.*	设备；器具
uninhabitable [ˌʌnɪn'hæbɪtəbl]	*adj.*	不适于居住的
furnishing ['fɜ:nɪʃɪŋ]	*n.*	家具；装饰品
newton ['nju:tən]	*n.*	牛顿（力的单位）
mass [mæs]	*n.*	[物理学]质量
slide [slaɪd]	*v.*	滑移，滑落；错动
adjacent [ə'dʒeɪsnt]	*adj.*	邻近的，毗邻的
equation [ɪ'kweɪʒn]	*n.*	方程式；等式

deformation [ˌdi:fɔ:'meɪʃn]	*n.*	变形
rod [rɒd]	*n.*	杆，拉杆，钢筋
stretching ['stretʃɪŋ]	*n.*	拉伸，伸长
elongating [i:'lɒŋgeɪtɪŋ]	*n.*	延长，加长
contraction [kən'trækʃn]	*n.*	收缩，缩减
deflection [dɪ' f lekʃn]	*n.*	变位，偏移
concave ['kɒn'keɪv]	*n.*	凹面，成凹形
dead load		恒荷载；静荷载
live load		活荷载
compressive force		压力
tensile force		拉力
shear force		剪力

Exercises

★ Ⅰ. Reading Comprehension

1. Describe the basic structural principles which must be observed in building construction.

2. What is dead load and what is live load?

3. Describe the tensile force and shear force.

4. Why does cracking in building occur?

5. What is called deflection?

★ Ⅱ. Translation

A. Translate the following sentences into English.

1. 在实体结构中，墙承受来自于屋面和楼面的荷载，并把荷载传递到基础。

2. 框架结构由于其墙体薄而轻，对高层建筑具有深远的意义，并促进了外部围护结构的发展。

3. 建造者必须考虑房屋是否能承受荷载、力、应力、应变。

4. 荷载下物体的变形称之为应变。拉力产生拉应变，压力产生压应变。

5. 当加在建筑材料上的应力超过其材料的强度，建筑就可能产生裂缝。

B. Translate the following paragraph(s) into Chinese.

Suspended concrete floors are generally of two types:

- in-situ concrete floors
- composite floor system

An in-situ concrete floor is one which is poured into formwork erected on the building site where the floor is actually required. The formwork is constructed in the shape of the required floor and the concrete is poured into it and left it to dry. The formwork is stripped away, leaving the concrete floor in place.

A composite floor can be a combination of steel decking and concrete or precast concrete floor.

In-situ reinforced slabs continue to be the most commonly used suspended floors because they are relatively economic and most builders are familiar with their construction.

★ Ⅲ. Group Activities

1. Work with your group members to summarize the advantages and disadvantages of steel framed structure, concrete framed structure and precast concrete framed structure.

2. Describe the basic in-situ concrete construction process to your partner.

The Pride of Chinese Architecture

扫一扫，听录音

Canton Tower is located at an intersection of Guangzhou New City Central Axis and Pearl River, facing Haixinsha Island and Zhujiang New Town directly in Guangzhou, Guangdong, China. It is 600 meters high, with 450 meters of its main body and 150 meters of its antenna,which makes it the highest tower in China and the third highest in the world. The original shape of Canton Tower is a beautiful and elegant lady with a slim figure. The uniqueness of the tower is not only embodied in its creative shape, but also owes to full considerations of aesthetics, material, structure, ergonomics and practical function.

Self-assessment

Now I know

- the differences between solid structure and skeletal structure are ______________;
- the advantages of framed structures are ______________;
- the different materials for framed structures are ______________;
- the advantages of reinforced concrete frames are ______________;
- the basic structural principles we should follow are ______________;
- the different types of building's loads are ______________;

Now I can

☐ use the correct terminology in the description of concrete construction;

☐ write a site report according to the workplace conversation;

☐ describe the materials and advantages of framed structure;

☐ explain different structural principles;

☐ explain the reasons for deformation and deflection.

Unit Five

Roof

Learning Objectives

After learning this unit, you will be able to

- write a site report according to the workplace conversation;
- explain the function and the performance requirements of roof;
- have basic ideas of flat roof and pitched roof;
- describe different types of roof strutting;
- use the correct terminology in the description of roof;
- understand Chinese excellent architectural culture.

Part One

Section A Workplace Conversation

扫一扫，听录音

Professional Words and Expressions

tiler ['taɪlə(r)]	*n.*	瓦工
ridge [rɪdʒ]	*n.*	屋脊
sagging ['sægɪŋ]	*n.*	下沉
truss [trʌs]	*n.*	桁架
foreman bricklayer		砌筑工工长，瓦工工长
roof trusses		屋架
party wall		分户墙；共用隔墙，界墙

It is a good site agent who can ensure that all work is properly supervised at all times. Small faults may go unnoticed until it is too late to avoid a lot of extra work.

★ Ⅰ. Listen to the conversation and fill in the blanks with what you hear.

(Peter—Site Agent; Jack—Foreman Bricklayer)

Peter: Hey, Jack, the tiler has just 1) ____________ this roof to me. Do you see anything wrong with it?

Jack: Oh yes, now you come to mention it. It looks as if the 2) ____________ is kind of sagging, doesn't it?

Peter: That's right. The tiler thought perhaps the 3) __________________ weren't strong enough or something. But as I told him, roof trusses always drop a little after tiling.

Jack: Hmm, so it's those 4) __________________, then. I suppose they've been taken up a bit too far. The trusses either side won't have been able to drop.

Peter: Exactly. And something has got to be done about it, you know. I must say, I think you ought to give those young 5) ____________ a few lessons. We can't have this sort of thing happening every time you take a week's holiday, can we?

Jack: Okay, Peter. I'll 6) _________________. Er—I'll get them to take twenty-five millimeters or so off those two party walls, and 7) _________________ all the rest up there.

Peter: Yes, that'll be all right. And I'll get the tiler to strip back the roof tiles where you'll need to 8) ____________________. Now, I want that roof looking decent by the end of the day.

Jack: Right. We'll see what we can do then.

★ Ⅱ. Pete writes a site report according to the conversation above.

This morning, the tiler pointed out the roof to me. It looks as if ____________

____

____

____

____

____

____

____

____

扫一扫，听录音

Section B Roof Construction

Roof is the most exposed part of the building to the weather. It is also the most important in keeping out the weather.

There are many types of roofing systems available and a careful choice is needed to suit the environment and also to give the desired design impact.

Roof can be defined as the upper covering of a building. Its most important function is to provide protection from the weather and the performance requirements are strength, stability, weather exclusion, thermal insulation, sound insulation, fire resistance, drainage, durability, lighting, ventilation, smoking dispersion and appearance. Therefore, roof structure should be strong enough to support the weight of coverings, its own weight, any imposed loads such as service equipment (chiller, water tank), and withstand wind loads. Roof structure can be timber (usually for pitched roof), metal (steel), or reinforced concrete. The larger the span, the lighter the roof and the imposed loads should be. Roof coverings are used to keep out the rain, wind and the sun.

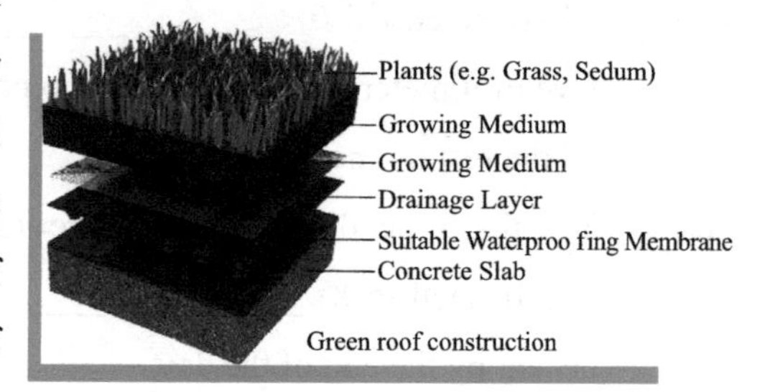

Green roof construction

Different types of roof coverings are available, for example, tiles/slates, metal sheets, waterproof membrane, leaves/straws and glass. Thermal insulation is to control heat loss and heat gain within the roof space, which is required in the local building regulation.

Generally, metal roofing and glass roofing are very poor in thermal insulation. The following methods of providing thermal insulation are applicable to all roof types:

· Using flexible or stiff insulating material in or under the roof cladding or structure.

· Using self-supporting insulating materials, such as wood wool slabs.

· Using lightweight concrete structure to give the required insulation.

· Building a secondary roof.

Sound insulation is important in noise sensitive buildings or in areas of high noise levels (e.g. near airports). Metal roofing is very poor in sound insulation. Therefore, suspended ceiling or double / triple layer of ceiling system may be required to improve sound insulation.

In roof construction, adequate resistance to fire is needed to prevent collapse of the roof before the evacuation of the occupants, and to prevent the spread of the fire from one part of building to the other.

Drainage is to be considered for the rainwater to be drained away in the simplest and most direct manner. Collecting rainwater for reuse may be one of the design features.

For durability, roof must be against weathering and atmosphere pollution, by which the metal roofing is prone to be attacked.

Flat roof may have more problems than pitched roof because of the rate of run-off, which may result in ponding. Lighting, ventilation and smoking dispersion may or may not be essential. Glazing can be used to allow for lighting. For large long building, smoke dispersion is important and roof vents are incorporated. Roof appearance is affected by degree of pitch of the roof surface, colour and texture of materials for roof coverings. Appearance is important for buildings such as mosque, churches, concert halls and buildings of importance.

There are many types of roofs of different shapes, pitch, materials and spans.

Flat roof is the one with its pitch less than 10°. The roof structure can be concrete, timber or metal, but outer covering must be waterproof. The structure is designed to carry the required loads. Reinforced concrete roof can accommodate services on it, or be used for garden and car park. The disadvantages of this type of roof are poorer appearance and drainage problem, and roof covering may not be durable.

Pitched roof has a 35° to 40°minimum pitch (domestic building). In the industrial-type, the pitch will be generally less. Coverings are usually in the form of corrugated metal or plastic sheets, tiles and slates. Structure is usually timber or steel sections. Metal roof coverings are mostly used in large span roof structures for they are light and thin to reduce deadweight. If they are stainless steel and copper, they can be durable and maintenance-free. Pitched roof is flexible in shape, suitable for dome and shell roof, which gives good aesthetics.

Services can be accommodated within the roof void (pitched roof) and on top of the roof for flat roof. The following services are often housed on flat roofs:

- lift motor rooms
- water storage tanks (for water supply and fire fighting)
- air-conditioning, plant-chillers, condensers
- exhaust fans

Roofs can be used for solar heating for both heating and hot water supply.

Professional Words and Expressions

performance [pə'fɔːməns]	*n.*	功能，性能；运行
strength [streŋθ]	*n.*	强度
thermal ['θəːməl]	*adj*	热的，保热的；温热的
insulation [ˌɪnsju'leɪʃn]	*n.*	绝缘或隔热的材料；隔声
resistance [rɪ'zɪstəns]	*n.*	抵抗；阻力；抗力；电阻
drainage ['dreɪnɪdʒ]	*n.*	排水
durability [ˌdjʊərə'bɪlətɪ]	*n.*	耐久性
dispersion [dis'pəːʃən]	*n.*	驱散；散布
chiller ['tʃɪlə]	*n.*	制冷机组
slate [sleɪt]	*n.*	石板瓦；石板
waterproof ['wɔːtəpruːf]	*adj*	防水的；用防水材料处理过的
flexible [fleksəbl]	*adj*	柔性的
stiff [stɪf]	*adj.*	刚性的
weathering ['weðərɪŋ]	*n.*	风化
ponding ['pɒndɪŋ]	*n.*	积水
glazing ['gleɪzɪŋ]	*n.*	门窗格玻璃；装（配）玻璃；镶嵌玻璃
steel section		型钢
condenser [kən'densə(r)]	*n.*	冷凝器
imposed load		活荷载，可变荷载；附加荷载
water tank		水箱

wind loads	风荷载
roof cladding	屋顶覆盖材料；屋面覆盖层
wood wool slabs	刨花板
lightweight concrete	轻质混凝土
suspended ceiling	吊顶；顶棚
domestic building	民用建筑
corrugated metal	金属波纹瓦；压型钢板
exhaust fans	排气扇

Exercises

★ Ⅰ. Answer the following questions.

1. What methods can be applied to provide thermal insulation to all roof types?
2. What roof structure should be chosen for sound insulation?
3. What are the disadvantages of flat roof?
4. Why are metal roof coverings used in large span pitched roof structures?
5. What services can often be housed on flat roofs?

★ Ⅱ. Match the following words or phrases with the correct Chinese.

1. thermal insulation	a. 隔热材料
2. weathering	b. 压型钢板
3. insulating material	c. 坡屋顶
4. corrugated metal	d. 隔热性
5. durability	e. 防水膜
6. fire resistance	f. 吊顶；顶棚
7. pitched roof	g. 风化
8. suspended ceiling	h. 耐久性
9. waterproof membrane	i. 民用建筑
10. domestic building	j. 阻燃性

★ Ⅲ. Read the following pictures.

1. Tell what each picture shows in a key word/ phrase from what you've learned in this unit.

1) ______________________

2) ______________________

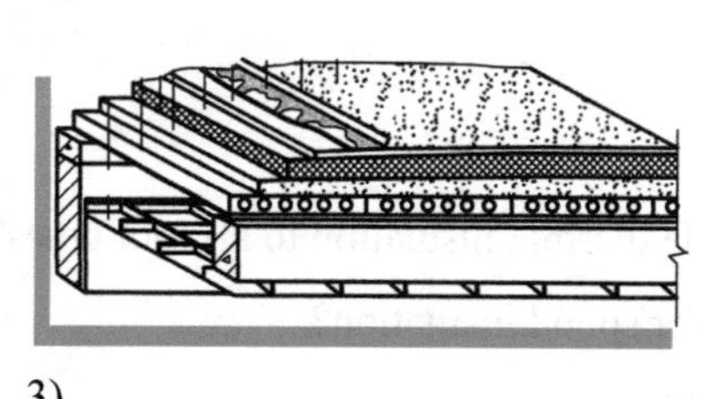

3) ______________________

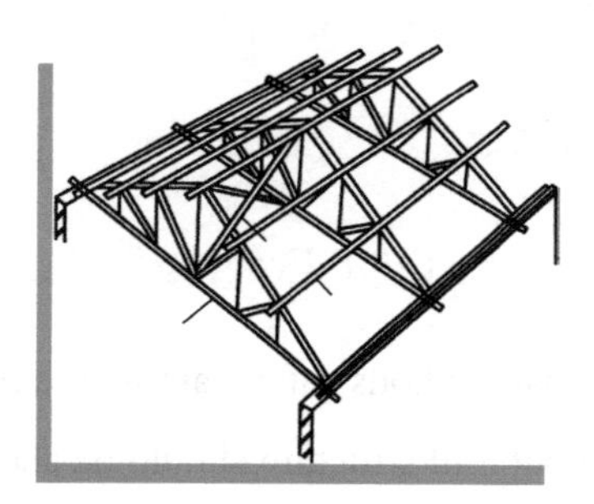

4) ______________________

5) ______________________

6) ______________________

2. Read the above pictures again, then tell your partners which pictures are related to each other, and what construction works these pictures show.

扫一扫，听录音

Part Two

Roof Strutting

A roof describes the portion of a structure intended to cover and give protection to the lower part of the building. Several factors influence the type of roof used, including the climate conditions, the plan of the building, span and materials used. Timber is still the most commonly used materials in some countries in the construction of domestic roof framing.

Roof strutting is a very important part of the roof frame, for if the roof is not strutted correctly, sagging will be very evident. The size, direction and fixings of all struts used for the support of roof members must be such that the loads from these struts or beams are transferred to the building foundations by the most direct route possible without causing deflections or lateral movement to the supporting structure.

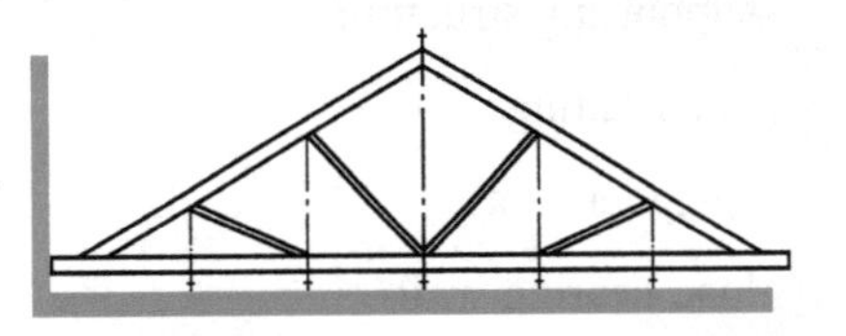

Single struts are fixed at an angle not exceeding 30° to the vertical when used perpendicular to the rafters or when used vertical to the rafters and running parallel with the underpurlin. When a single strut exceeds 30° but does not exceed 45° the strut shall be opposed by another strut at the same angle. This type of roof strutting is known as fan strutting and must have one spreader cleat bolted to each side of the struts to help maintain stability.

Struts are to be birdsmouthed or halved to the underpurlins, and be supported only on load bearing walls or adequate strutting beams. Where supported by stud wall, the strut must sit immediately over a stud or their load must be adequately distributed to two or more studs by intermediate blocking. Struts supporting underpurlin over openings should be avoided, but where no alternative can be found, the lintel above such an opening must be designed to accommodate the additional load.

Professional Words and Expressions

strut [strʌt]	*n.*	对角撑；撑杆
fixings ['fɪksɪŋz]	*n.*	紧固件
members ['membə(r)]	*n.*	构件
perpendicular [ˌpɜ:pən'dɪkjələ(r)]	*adj.*	垂直的；垂直
rafter ['rɑ:ftə(r)]	*n.*	椽条；椽子
purlin ['pɜ:lɪn]	*n.*	檩条

bolt [bəʊlt]	*v.*	拴住；螺栓
birdsmouth [bɜ:dz'maʊð]	*v.*	开齿槽；角口承接，企口结合
halve [hɑ:v]	*v.*	半对搭；相嵌接合
stud [stʌd]	*n.*	龙骨；立筋
lintel ['lɪntl]	*n.*	过梁
roof strutting		屋顶支撑
supporting structure		支承结构；固定架
fan strutting		扇形撑
spreader cleat		加固托座
load bearing wall		承重墙
strutting beams		支撑梁
stud wall		立筋隔墙；立柱墙

Exercises

★ Ⅰ. Reading Comprehension

1. What main factors influence the type of roof used?
2. Why is roof strutting a very important part of the roof frame?
3. In order to support roof members, what requirements should struts meet?
4. What should be done to roof strutting with a single strut ranging from 30° to 45°?
5. How can struts be set to reasonably distribute the loads ?

★ Ⅱ. Translation

A. Translate the following sentences into English.

1. 屋顶是对房屋下部起覆盖和保护作用的围护结构。
2. 屋顶一般分为平屋顶和坡屋顶两种。
3. 在有些国家，修建民用建筑的屋顶通常选用木材。
4. 与坡屋顶相比，平屋顶的主要优点是便于施工，造价较低。
5. 为了利于排水，平屋顶往往也会有一定的坡度。

B. Translate the following paragraphinto Chinese.

In the design of pitched roofs, one of the most important factors is the degree of the pitch or slope, which depends mainly on the material used to cover the roofs. The steeper the pitch, the more effective the roof is in quickly disposing of rainwater or snow. On the other hand, a steeper pitch entails a larger roof area and a higher cost.

The roof space, commonly known as a loft, which is quite large in a roof with a steep pitch, is often used to provide additional rooms or a storage area.

★ Ⅲ. Group Activities

1. What materials are mainly used to construct house roof? What advantages and disadvantages do they have respectively?

2. What is the best house roof in your mind? Why?

The Pride of Chinese Architecture

扫一扫，听录音

Yellow Crane Tower, standing atop Snake Hill, is the most famous attraction in Wuhan. It was first built during the Three-Kingdom Period in 223 CE as a military watchtower and was destroyed several times in history. The existing 5-layer Yellow Crane Tower was built in 1985 with a height of 51.4 meters. The roof is covered by 100,000 yellow glazed tiles. With yellow upturned eaves, each floor of the tower looks like a yellow crane spreading its wings to fly. As many famous poets and celebrities have been to the Yellow Crane Tower and left many popular poems about it, it becomes very famous in China and is listed among the three greatest towers south of the Yangtze River.

Self-assessment

Now I know

- the most important function of roof is ______;
- the performance requirements for roof are ______;
- the methods of providing thermal insulation to all roofs are ______;
- the advantages and disadvantages of flat roof are ______;
- the features of pitched roof are ______;
- the services often housed on flat roofs are ______;
- the requirements for the size, direction and fixings of all roof struts are ______.

Now I can

- ☐ use the correct terminology in the description of roof;
- ☐ write a site report according to the workplace conversation;
- ☐ explain the function and performance requirements of roof;
- ☐ have basic ideas of flat roof and pitched roof;
- ☐ describe different types of roof strutting.

Unit Six

Walls

Learning Objectives

After learning this unit, you will be able to

- write a site report according to the workplace conversation;
- describe the way of laying bricks;
- know the various types of walls commonly used;
- explain the load bearing walls and non-load bearing walls;
- use the correct terminology in the description of wall construction;
- understand Chinese excellent architectural culture.

Part One

Section A Workplace Conversation

扫一扫，听录音

Professional Words and Expressions

brickwork ['brɪkwɜ:k]	*n.*	砖砌体；砌砖，砌砖工程
frog [frɒg]	*n.*	（砖面）凹槽
refer [rɪ'fɜ:(r)]	*v.*	送交
sound reduction		隔声

Which way up does a brick go?

One of Jack's bricklayers thinks it makes no difference which way up a brick is laid. But the architect who wrote the specification disagrees, and so does the clerk of works.

★ Ⅰ. Listen to the conversation and fill in the blanks with what you hear.

(Jack—Foreman Bricklayer; Sid—Clerk of Works; Peter—Site Agent; Mark—Bricklayer)

Jack: Hello, Sid. Peter said you weren't very happy about this 1)__________. He asked me to sort it out with you.

Sid: Yes, that's right. Young Mark started that wall this morning, and he's laid every brick 2)__________down as far as I can see.

Jack: Oh dear, I'm sorry. Well, I'll have a word with them about it. I suppose it could have been worse, couldn't it?

Sid: Well, maybe it could, but now it doesn't comply with the 3)__________. I told him earlier that he should have to take it down. He took no notice then-and he wasn't very polite either. I'm really annoyed about it.

Jack: Well, be reasonable, Sid, I'll see he does the rest of it properly.

Sid: That doesn't make any difference. It'll still have to be rebuilt, you know. There'll be 4)__________ in those frogs and there are unfilled joints everywhere. The wall just won't have adequate 5)__________.

Jack: I know, I know. But surely that doesn't matter in this case, does it? It's only going to be a 6)__________, after all.

Sid: Well, that's not really the point. It's 7)__________, and I don't want to see it happen again where it will matter.

Jack: Well, I'm sorry, but I think you're being unreasonable.

Sid: Okay, then. If that's the way you feel, I'll have to 8)__________ for his decision. You can't say that's unreasonable.

★ Ⅱ. Peter writes a site report according to the conversation above.

This morning, Sid wasn't happy about the brickwork. I asked Jack to sort it out with him.____

____

____

____

____

____

____

Section B Wall Constructions

扫一扫，听录音

A wall is a vertical element to enclose the building from the external environment, or enclose the space within the building. It may also divide the space.

There is a wide range of materials available for wall constructions such as brick, block, stone, reinforced concrete and timber, and a careful selection is needed bearing in mind the requirement to be met.

Walls can be classified as load bearing walls which support loads from floors and roof in addition to its own weight, nonbearing walls which carry no floor or roof loads except its own weight, external non-load bearing walls for framed structure, partitions which can be load bearing or non-load bearing to divide space into rooms, separating walls which are internal walls to separate adjoining buildings, compartment walls which divide space into compartments for purpose of fire protection, and retaining walls which support and resist the fall of the soil.

The functions for the external wall are to enclose the structural frame or building to exclude weather and to provide an aesthetic appearance. The functions for the internal wall are to divide the space within the building. Load bearing walls are required to have certain strength and stability to be able to resist stresses due to its own weight, superimposed loads and lateral pressure such as wind. Walls are also required to be stable to avoid overturning and buckling due to excessive slenderness. Walls must be able to support building services, equipment and plant such as air conditioning unit, sanitary appliances and hand drier which may be heavy. Lightweight wall, glass block wall may not be suitable for those loads. Walls need weather resistance to avoid rain and wind penetration. Bathroom-wall should be damp resistant. Walls must provide a barrier to heat gain or heat loss which increases the cost of cooling or heating in order to maintain a comfortable internal environment. Plant room and boiler room generate a lot of heat where louvers/ louvered wall may be used to remove it. For noise sensitive buildings such as hospitals, schools, concert halls, theatres, etc. Thicker and heavier walls are required for sound insulation to exclude noise from traffic, aircraft, train, building services, plant and equipment, with window and door openings carefully sealed. In special circumstances, such as theatres and studios, the walls need to be sound absorbing to reduce the reflections of sound within the room, which can be achieved by lining the walls with porous sound absorbers or thin panels placed in front of the wall surface. Walls should be fire resistance to prevent spread of fire, with fire-stops to the openings and penetrations, fire-resisting materials to plant room, lift shaft, and fire escape staircase.

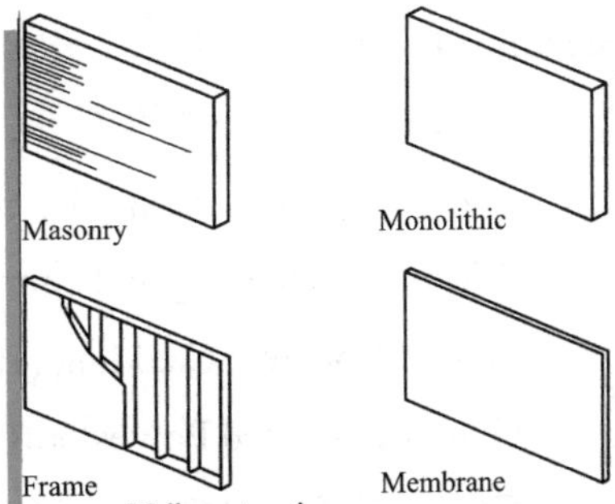

Masonry Monolithic Frame Membrane

Wall construction

Different types of walls are used in construction. Masonry wall is build by individual brick/ blocks cemented together. It may be either a solid masonry wall or a masonry cavity wall with either reinforced or unreinforced. Monolithic wall is a concrete wall which may be either plain or reinforced. Frame wall is made of timber or metal with facing/board on each side to form a completed system. Membrane wall is just like sandwich of two thin skins of plastic, metal, plywood bonded to a core.

External wall can be constructed of bricks laid in mortar and overlapped in some form of bonding. The advantages of brick wall are good fire resistance and moderately good thermal insulator. Brick wall does not deteriorate structurally and requires very little maintenance over a long period of time. Plastered both sides, it should give acceptable sound insulation against external noise sources. Block wall is extensively used for both bearing and non-load bearing walls. Blocks can be hollow clay, hollow concrete, solid concrete or light weight concrete. Block wall constructs stretcher bond only as there is no need to bond into the thickness of the wall. It has advantages of brick wall except for sound insulation and requirement of finishes. Reinforced concrete wall is mainly used for basement walls, lift shafts, retaining walls, and services ducts for its fairly watertight and good fire- resistance. The disadvantages of it are low thermal insulation, poor wall surface which requires finishing or facing and cracks due to settlement especially for unreinforced walls. In fill panel which is non-load bearing is fixed between main structural members to fulfill the functional requirements of an external wall, usually timber, glass, or metal. Cladding and curtain walling systems are elements attached to structural frame by spanning between given points of supports. They carry no load except their own weight but should be able to resist wind and rain. The designs of the systems are more interesting and challenging for their better appearance, different colours and configuration.

Notes:

1. stretcher bond 顺砖砌合法；其他砌体结构的砌合法如：English bond 英式砌合法（顺丁分层砌合）；flemish bond 荷兰式砌合法（同层丁顺砖交错）；

common bond 普通砌合法 / 美式砌合法（多顺一丁砌合）

stretcher（顺砖，顺砌砖），header（丁砖，丁砌砖）

soldier（立砖，立砌砖），rowlock（竖砖，竖砌砖）

2. Types of cladding and curtain walling systems (围护结构及幕墙结构的种类):

Precast Concrete Cladding (50 mm thick) 预制混凝土骨架外墙板

Glass Reinforced Polyester (GRP) (1-3mm) 玻璃纤维增强聚酯层

Glass Reinforced Cement (GRC) 玻璃纤维增强混凝土

Curtain Walling (glass/ metal panels) 幕墙（玻璃 / 金属板）

Professional Words and Expressions

partition [pɑ:'tɪʃn]	*n.*	隔墙，隔断
compartment [kəm'pɑ:tmənt]	*n.*	间隔，隔断；分隔间
superimposed [sju:pərɪm'pəʊzd]	*adj.*	附加的，叠加的
lateral ['lætərəl]	*adj.*	侧面的，横向的
overturn [ˌəʊvə'tɜ:n]	*v.*	垮，倾覆
buckle ['bʌkl]	*v.*	变形，弯曲
buckling		失稳
slenderness ['slɛndɚnɪs]	*n.*	长细比；细长度
penetration [ˌpenɪ'treɪʃn]	*n.*	渗透，穿透
barrier ['bærɪə(r)]	*n.*	屏障；隔离，障碍
louver ['lu:və]	*n.*	百叶（式），格栅；百叶窗
line ['laɪn]	*v.*	给……加内衬；排列，衬砌
porous ['pɔ:rəs]	*adj.*	多孔的，能渗透的
absorber [əb'sɔ:bə]	*n.*	吸收器，吸收装置
sound absorber		吸声材料
panel ['pænl]	*n.*	板，镶板，嵌板
mortar ['mɔ:tə(r)]	*n.*	砂浆
overlap [ˌəʊvə'læp]	*v.*	搭接，重叠
bond [bɒnd]	*v.*	砌合，结合，组合
	n.	组砌法；结合（物），黏合（剂）
deteriorate [dɪ'tɪərɪəreɪt]	*v.*	变质，老化，退化
maintenance ['meɪntənəns]	*n*	维修，保养

reinforced concrete	钢筋混凝土
compartment wall	防火墙
separating wall	隔墙；分户墙
hand drier	干手器
glass block	玻璃砖
fire-stop	阻火材料，阻火部件
masonry wall	砌体墙
monolithic wall	实心墙
membrane wall	夹芯墙
stretcher bond	顺砖砌合；全顺组砌法
infill panel	内镶（嵌）板

Exercises

★ I. Answer the following questions.

1. What types of materials can be used for wall construction?
2. What are the functions of walls?
3. What are the advantages of brick walls?
4. How is the block wall constructed and what are the advantages of it?
5. Where are the reinforced concrete walls mainly used?

★ II. Match the following words or phrases with the correct Chinese.

1. superimposed load	a. 隔板墙
2. membrane wall	b. 挡土墙
3. configuration	c. 整体墙
4. stretcher bond	d. 隔墙
5. penetration	e. 附加荷载
6. monolithic wall	f. 结构，配置
7. mortar	g. 顺砖砌合
8. retaining wall	h. 渗透
9. separating walls	i. 空心混凝土
10. hollow concrete	j. 砂浆

★ Ⅲ. Read the following pictures and tell the class what you know.

1. Find a proper word or phrase to describe each of the following pictures.

1) ______________________________  2) ______________________________

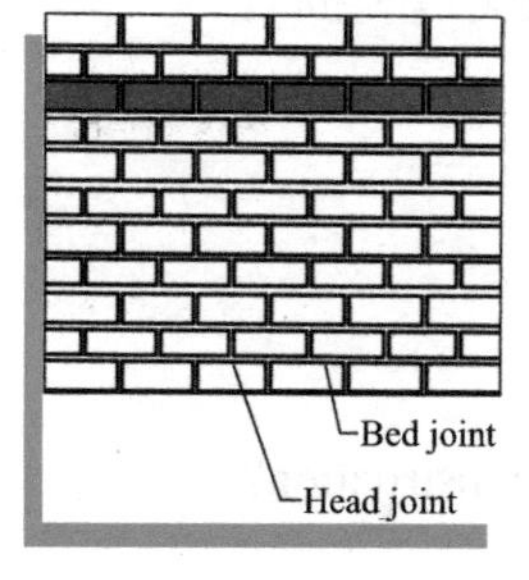

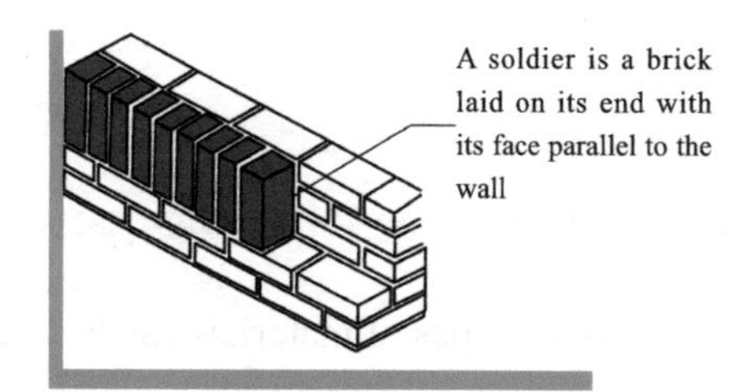

3) ______________________________  4) ______________________________

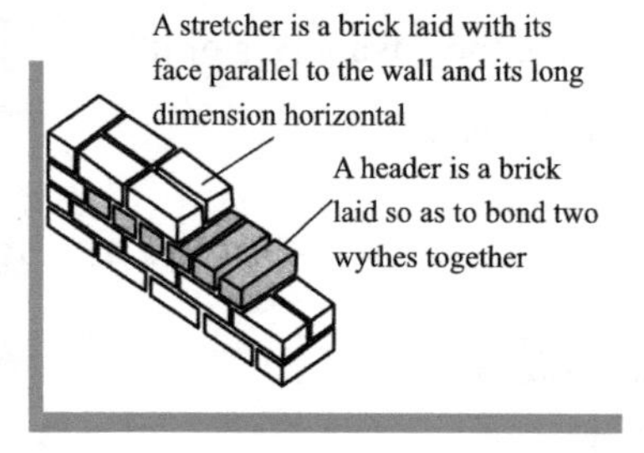

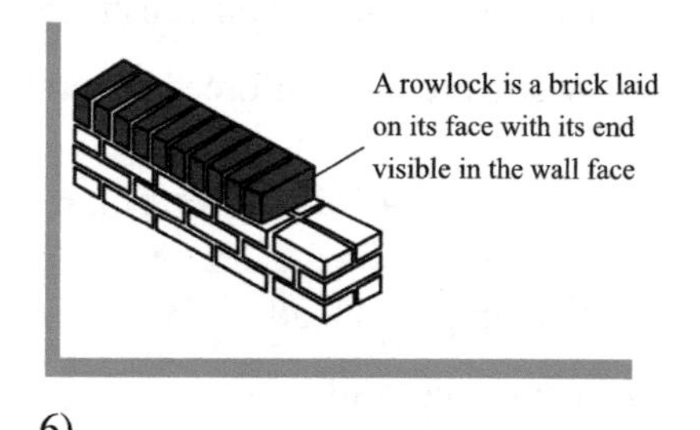

5) ______________________________ 6) ______________________________

2. Read the pictures again and tell your partners what kind of construction work they show and how to do the work.

扫一扫，听录音

Part Two

Load Bearing Walls

Walls that carry a weight other than their own are known as load bearing walls and are an important component of the structure of a building. The crucial factors to be considered when designing a load bearing wall include:

· the condition of loading,

· the conditions of the vertical and lateral support,

· the allowable stresses (or load factor),

· whether there are any doors, windows or other openings in the wall.

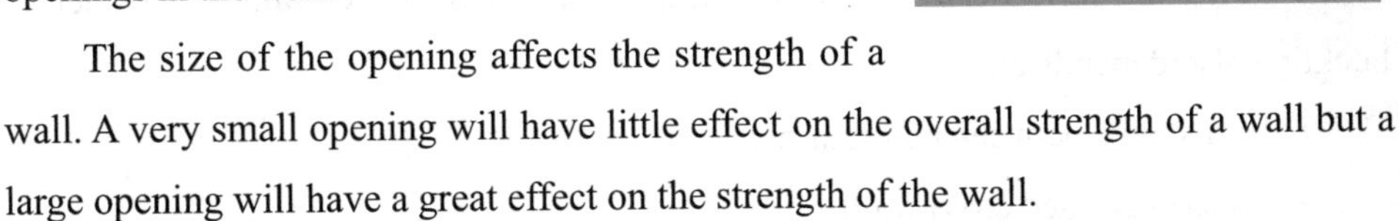

The size of the opening affects the strength of a wall. A very small opening will have little effect on the overall strength of a wall but a large opening will have a great effect on the strength of the wall.

A load bearing wall should be constructed from concrete, wood, masonry or metal in different forms.

Loads and other forces and actions which affect the stability of a masonry wall will include:

· dead loads (the mass of building components),

· live loads (including wind load),

· long term dimensional changes of brickwork (such as expansion of clay bricks),

· movement of structural members (including shrinkage, expansion, deflection or rotation of other components such as concrete floors, roof frames, structural elements),

· foundation movement.

It is important for you to understand how walls are constructed to enable them to withstand the effects of lateral (in the direction of thickness) and vertical forces placed on them.

Lateral Loads on Walls

In lightly loaded buildings, the lateral loads from strong winds are possible more of a problem than vertical loads. The main consideration is to stop the wall from overturning. This overturning failure caused by wind pressure transmitted directly to the external walls can be prevented by:

· providing intersecting walls,

· anchoring timber framed roofs against uplift, which also provides lateral support.

Long walls may also fail under lateral load by the bricks turning on their bed joints. This type of failure is resisted by the mortar joint between the masonry units. Because of this, bond strength is often more important than crushing strength.

Vertical Loads on Walls

Tall thin masonry walls tested under eccentric or off centre loads will generally fail by buckling.

The intensity of the load at which this occurs will depend on the slenderness of the wall—the ratio between the height and thickness of the wall. Resistance to buckling also depends on:

· the distance between end returns,

· properly bonded cross walls,

· the amount of lateral support a wall receives from floor and roof systems.

Brick Walls Construction

Brickwork should be laid accurately, plumb and bonded. Bed joints (horizontal) and perpend joints (vertical) are not to exceed 10mm in thickness. Bricks should be laid on a full bed of mortar with perpend joints completely filled. Before commencing brickwork, the bricklayer should be informed by the builder / supervisor of the position of all damp-proof material and flashings. On completion of bricklaying, all mortar splashes and stains should be removed and the brickwork left in a clean finished condition.

Professional Words and Expressions

opening ['əʊpnɪŋ]	*n.*	洞口；开口
shrinkage ['ʃrɪŋkɪdʒ]	*n.*	收缩
expansion [ɪk'spænʃn]	*n*	膨胀
rotation [rəʊ'teɪʃn]	*n.*	旋转，扭转
uplift ['ʌplɪft]	*n.*	拔起；提起；隆起
eccentric [ɪk'sentrɪk]	*adj.*	偏心的
dimensional change		尺寸变化
intersecting wall		相交墙，交叉墙
bed joint		水平缝，层间接缝
mortar joint		灰缝，砂浆接缝
eccentric load		偏心荷载
end return		端墙（return 迂回墙）

Exercises

★ Ⅰ. Reading Comprehension

1. What is a load bearing wall?

2. What factors should be taken into consideration when we build a load bearing wall?

3. What are the two main types of loads on walls?

4. What factors can affect the stability of a masonry wall?

5. How should brickwork be laid?

★ Ⅱ. Translation

A. Translate the following sentences into English.

1. 承重墙是指支撑着上面楼层重量的墙体，如果打掉，整个建筑结构会被破坏。

2. 墙上一个非常小的洞口对墙体的整体承受力影响很小，但大的洞口却有很大的影响。

3. 承重墙应用不同形式的混凝土、木材、砖石或金属建造。

4. 医院、学校、剧院等对噪声敏感场所的建筑，需要选用较厚重的墙来达到隔声的效果。

5. 承重墙是经过科学计算的，如果在承重墙上开洞装修，就会影响结构的稳定性。

B. Translate the following paragraph(s) into Chinese.

Bricks are laid in the various positions for visual reasons, structural reasons, or both. The simplest brick wall is a single wythe（砖厚）of stretchers（顺砖）. For walls two or more wythes thick, headers（丁砖）are used to bond the wythes together into a structural unit. Rowlock（竖砖）courses are often used for caps on garden walls and for sloping sills under windows, although such caps and sills are not durable in severe climates. Architects frequently employ soldier（立砖）courses for visual emphasis in such locations as window lintels or tops of walls.

★ Ⅲ. Group Activities

1. What are the differences between a load bearing wall and a non-load bearing wall?

2. What should we bear in mind when we decorate the part of a house supported by a load bearing wall?

扫一扫，听录音

The Pride of Chinese Architecture

The Great Wall of China, the largest man-made project in the world, is a series of ancient fortifications built in northern China with about 6700 kilometers from east to west. It is one of the Eight Wonders of the world and was listed as a World Heritage by UNESCO in 1987.

The construction of the Great Wall began from the Warring States period (7th century BC) and it was continuously built to the 17th century AD as military defence project of successive Chinese Empires. Although named the "wall", it is an integrated system including not only solid walls, but also massive beacon towers, barriers and fortresses (堡垒), which made it the world's largest military structure.

Self-assessment

Now I know

· the function of a wall is to______________________________;

· the materials for walls can be______________________________;

· the different types of walls are______________________________;

· walls should meet the requirements of ________________________;

· the factors that affect the stability of masonry walls are ________________;

· special attention should be paid to ________________________ in brick walls construction.

Now I can

☐ use the correct terminology in the description of wall construction;

☐ write a site report according to the workplace conversation;

☐ describe different types of walls and their functions;

☐ explain different materials of walls and their advantages;

☐ explain the difference between load bearing wall and non-load bearing wall.

Unit Seven

Floor Construction

Learning Objectives

After learning this unit, you will be able to

- know different types of floor construction;
- write a site report according to the workplace conversation;
- know the functional requirements of floor;
- know the construction of concrete floor on fill;
- learn to use the correct terminology in the description of floor construction;
- understand Chinese excellent architectural culture.

Part One

Section A Workplace Conversation

扫一扫，听录音

Professional Words and Expressions

tile [tail]	*n.*	瓷砖；面砖；瓦
screed [skri:d]	*n.*	砂浆底层，找平层
crack ['kræk]	*v.*	开裂
level ['levl]	*v.*	拉平；找平
grout [graʊt]	*v.*	勾缝，灌浆
bed [bed]	*v.*	铺砖；将……铺平
semi-dry method		半干法
separating layer		分离层
polythene sheeting		聚乙烯护板
differential movement		不均匀变形

Floor Finishes

You do it that way. I do it this way.

There are more ways than one of doing some jobs, and each tradesman chooses to use the method which he has found to be the quickest and the best.

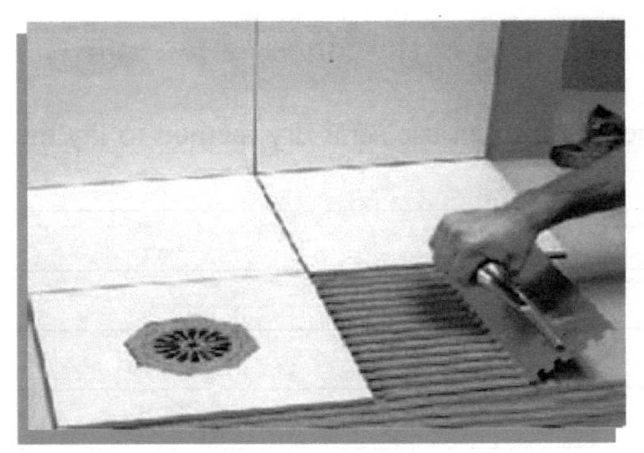

★ **I. Listen to the conversation and fill in the blanks with what you hear.**

(George—Foreman Tiler; Dave—Tiler)

Dave: I don't often use this semi-dry method to lay the tiles. It seems to take me a long time. Do you always do it this way?

George: Oh yes. I've usually found it 1)____________. What do you think of it?

Dave: Well, I can't say I'm very keen on it. Have you ever tried any other methods? On my last job we used a 2)____________ — you know, we laid the tiles on polythene sheeting over a screed. It seemed to work really well.

George: Hmm, yes, I've sometimes done it that way myself.

Dave: We never got any 3)____________ because of differential movement that way.

George: Well, my way either. Besides, with your method there are two parts to the job. First you've got to level your screed accurately, and then 4)____________ accurately afterwards, haven't you? That all 5)____________.

Dave: Well, your method has two stages too, if you think about it. You've got to make all your semi-dry mix first, and get it laid, and then you've got to start mixing your 6)____________ to bed the tiles on. And you've still got to fill and grout the joints afterwards with some of each mix.

George: But listen, if you're laying screeds, you've always got to leave them to 7)____________, haven't you? It's a waste of time, which is when you're working on a small area.

Dave: I don't know about that.

George: Well, if we compare the two methods next week, mine will be quicker ninety percent of the time. You just see if I'm right.

Dave: Okay, George, you win. We'll 8)__________________________.

★ Ⅱ. George writes a site report according to the conversation above.

This morning, Dave told me that he didn't often use semi-dry method to lay the tiles. ____
____
____
____
____
____

Section B　Floor Construction

扫一扫，听录音

Floors are horizontal structural elements. There are different types of floor constructions for different loading and span requirements. A careful choice is to be made.

Concrete is the main material used for the construction of floors. Whilst in-situ concrete floors are widely used, precast floor constructions offer advantages that can outweigh the main disadvantage of being more costly.

Floors are constructed for comfortable surface for use, to support live loads such as occupants, machines and furniture, and to divide the space vertically to increase floor area. The functional requirements for floors are: 1) strength and stability which are able to support its own weight and imposed loads, weight of building services equipment & plant; 2) thermal insulation which are important in cold countries or areas and can be achieved by using thermal insulation layer or hollow floor construction; 3) noise and vibration control which needs correct choice of floor types, the thicker the floor the better the insulation; 4) fire resistance which are especially important for upper floors, plant room and boiler room; 5) damp penetration resistance which are important for substructure such as basement, lift pit & ground floor, and upper floor only on bathroom and kitchen area, where waterproofing membrane must be used; 6) durability which means to be able to withstand for a period against water, fire and wear and tear by correct choice of floor finish; 7) accommodation of services which are required in the floor to provide easy access; 8) provision of acceptable surface finish which is particularly for bathroom, kitchen, and heavy wear and tear such as canteen and industrial building.

Floors can be classified as solid floor and suspended floor. The former is at ground and basement levels, resting directly on the ground with materials of plain or reinforced concrete. The thickness of the floor depends on the loads and bearing capacity of the soil. The latter refers to upper floor which spans between two supports. The suspended floor may be timber, reinforced concrete or steel.

Solid ground floor comprises of several layers as it is supported directly by the ground. The construction details are as follows:

Hardcore

· it comprises broken bricks, stone,

· about 150 mm thick,

· well compacted,

· to ensure a consistent material over the whole area so that the loading of the floor is uniformly spread over the area,

· to reduce capillary action of moisture from the ground,

· to raise the finish surface above ground level,

· to provide a level surface for laying of DPM(Damp Proof Membrane),

· to provide a clean, dry and firm working surface.

Concrete Blinding

· it consists of a layer of fine ash or sand 25-50mm thick or a 50-75 mm layer weak concrete,

· to fill the voids in the hardcore in the surface region,

· to provide a level surface for DPM and prevent it from being punctured by hardcore,

· to provide a true surface from which the reinforcement can be positioned.

Damp Proof Membrane (DPM)

· waterproofing layer,

· can be placed either above or below the slab,

· DPM should overlap with DPC in the surrounding walls,

· polythene sheet, bitumen, asphalt.

Concrete Slab

· to support loads,

· between 100 and 150mm thick.

Wearing Surface

· usually takes the form of a cement ： sand screed 1 ： 3 mix,

· thickness varies from 25 to 65mm,

· take out any surface irregularities in the concrete bed.

Suspended concrete floors are generally of two types:

· in-site concrete floors,

· composite floor systems.

An in-situ concrete floor is one which is poured into formwork erected on the building site where the floor is actually required. The formwork is constructed in the shape of the required floor and the concrete is poured into it and left to dry. The formwork is stripped away, leaving the concrete floor in place.

A composite floor can be a combination of steel decking and concrete, or a precast concrete floor.

For most applications, precast slab elements are manufactured with a rough top surface. After the elements have been erected, a concrete topping is poured over them and finished to a smooth surface. The topping, usually 2 inches (50mm) in thickness, bonds during curing to the rough top of the precast elements and becomes a working part of their structural action. The topping also helps the precast elements to act together as a structural unit rather than as individual planks in resisting concentrated loads and diaphragm loads, and conceals the slight differences in camber that often occur in prestressed components. Structural continuity across a number of spans can be achieved by casting reinforcing bars into the topping over the supporting beams or walls.

Professional Words and Expressions

damp [dæmp]	*adj.*	潮湿的
capillary [kə'pɪlərɪ]	*adj.*	毛发状的，毛细作用
hardcore ['hɑ:dkɔ:]	*n.*	碎砖垫层，硬石填料
puncture ['pʌŋktʃə(r)]	*v.*	刺穿
bitumen [bɪ'tu:mən]	*n.*	沥青，柏油
asphalt ['æsfælt]	*n.*	（铺路等用的）沥青混合料
irregularity [ɪˌregjə'lærətɪ]	*n.*	不规则；不整齐
composite ['kɒmpəzɪt]	*n.*	合成物，混合物，复合材料
strip [strɪp]	*v.*	除去，剥去；拆模
topping ['tɒpɪŋ]	*n.*	面层，覆盖层

plank [plæŋk]	*n.*	板，支持物；厚木板
conceal [kən'si:l]	*v.*	隐藏，遮住
camber ['kæmbə(r)]	*n.*	拱形，弧形
wear and tear		磨损，损坏，损耗
DPM (Damp Proof Membrane)		防水膜
DPC (Damp Proof Course)		防水（潮）层
wearing surface		耐磨面；磨耗面，磨损面
steel decking		（压型）钢板，钢衬板；钢板层
diaphragm load		隔板荷载

Exercises

★ Ⅰ. Answer the following questions.

1. What material is most widely used for constructing floor?
2. What are the functional requirements for floors?
3. How is the solid ground floor constructed?
4. How is the in-situ concrete floor constructed?
5. What is topping? And why is it important for precast concrete floor?

★ Ⅱ. Match the following words or phrases with the correct Chinese.

1. asphalt	a. 碎砖垫层
2. plank	b. 聚乙烯护板
3. steel decking	c. 厚板，木板
4. hardcore	d. 钢板层
5. DPC	e. 压紧，压实
6. formwork	f. 防水层
7. imposed load	g. 沥青
8. polythene sheet	h. 顶部结构
9. compact	i. 模板
10.topping	j. 外加荷载

★ Ⅲ. Read the following pictures.

1. Find a proper word or phrase to describe each of the pictures.

1) ______________________________ 2) ______________________________

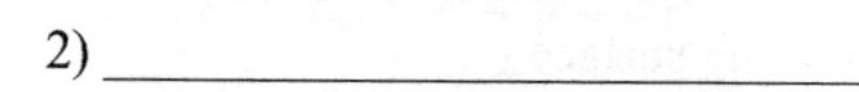

3) ______________________________ 4) ______________________________

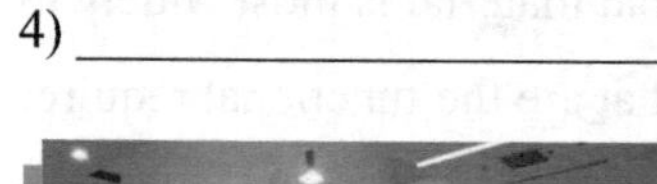

5) ______________________________ 6) ______________________________

2. Read the pictures again and tell your partners what the function is for each picture.

扫一扫，听录音

Part Two

Concrete Floors on Fill

There are two types of concrete floor systems poured on filling:

· slabs poured on fill to support the walls of a structure (raft slab),

· floors poured between internal masonry walls (floating slab).

The most important part of the pouring of a raft slab is the preparation of the compacted fill used to support it. The structural stability will depend on how well this base has been prepared.

A floating slab varies from a raft slab in that it is generally constructed without load bearing edge beams and within the perimeter of masonry walls. In floating slab construction the floor slab rests directly on a stabilized base, generally of compacted sand, with a moisture proof membrane directly under the slab.

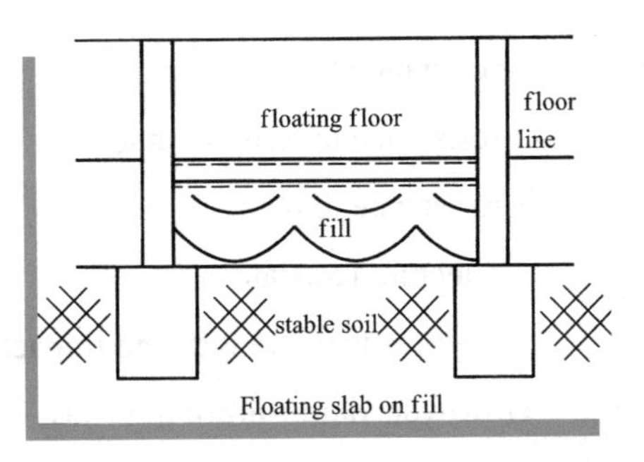

Floating slab on fill

Floating slabs are generally used in the wet areas in a cavity brick building that has timber floors in the main rooms.

Note that where floating slabs are used:

- The slab should be poured on consolidated (hardcore) filling.
- The filling should be clean and free from clay, cans, broken bricks, vegetation etc.
- The slab should be clear of the footings.

Thickness and Reinforcement

All concrete floors on fill should be reinforced with fabric in accordance with specification, and be placed 25 mm down from the top of the slab for crack control.

Verandah and porch concrete floors slabs must be at least 75 mm thick reinforced concrete and all other concrete floors at least 100 mm thick.

Damp Proof Course

A damp proof course should be built into walls adjoining in fill floor slabs on membranes. It should be placed so that it is above the underside of the slab in internal walls and inner leaves of cavity walls, project 40 mm and run down over the under floor membrane turned up against the wall.

The requirements of the damp proofing of floor on the ground are as follows.

If a floor of a room is laid on the ground or on fill, moisture from the ground must be prevented from reaching the upper surface of the floor and adjacent walls by the insertion of a vapour barrier that has following qualities:

- impermeability,
- durability,

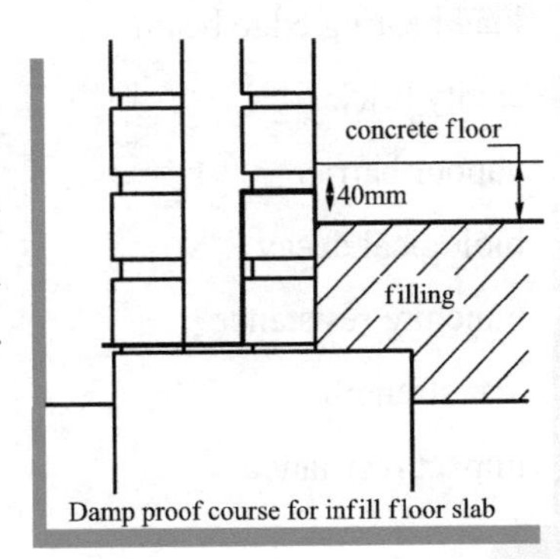

Damp proof course for infill floor slab

- resistance to biological decay,
- resistance to termite attack,
- resistance to damage,
- puncture resistance,
- requires tear strength and impact resistance.

However, damp-proofing may not be provided if the building is exempt from weatherproofing, e.g. garages, tool shed, non-habitable out buildings, or the floor is the base of a stair, lift or similar shaft which is adequately drained by gravitation or mechanical means.

Professional Words and Expressions

pouring [pɔːrɪŋ]	*n.*	浇筑
fill [fɪl]	*n.*	填土，回填料
moisture ['mɒɪstʃə]	*n.*	潮气；水分，湿度
vegetation [vedʒɪ'teɪʃ(ə)n]	*n.*	植被
verandah [və'rændə]	*n.*	走廊
porch [pɔːtʃ]	*n.*	走廊，门廊
impermeability [ɪm,pɜːmɪə'bɪlətɪ]	*n.*	抗渗性
termite ['tɜːmaɪt]	*n.*	白蚁
drain [dreɪn]	*v.*	排水
mechanical [mɪ'kænɪk(ə)l]	*adj.*	机械的，设备的，力学的
concrete floor		混凝土底板
raft slab		筏板，筏形基础
floating slab		浮板，悬浮板
load bearing edge beam		承重边梁
cavity brick		空心砖
vapour barrier		隔汽层
biological decay		生物分解
puncture resistance		抗穿刺性，抗冲击
tear strength		撕裂强度
impact resistance		抗冲击性

Exercise

★ Ⅰ. Reading Comprehension

1. What are the two types of concrete floor systems?

2. Why should we make good preparation of the base when constructing the floor?

3. How do we distinguish a floating slab from a raft slab?

4. What is the place where floating slabs are used like?

5. How do we build damp proof course?

★ Ⅱ. Translation

A. Translate the following sentences into English.

1. 地面基层的状况决定了筏板浇筑结构的稳定性。

2. 混凝土地面应按照规范进行加固。

3. 防潮层是防止地下潮气上升至地面、墙面以及室内空间中的结构。

4. 阳台和门廊的楼板必须至少为 75mm 厚的钢筋混凝土。

5. 浮板一般都用在潮湿的地方，主要是用于木地板房间。

B. Translate the following paragraph(s) into Chinese.

An in-situ concrete slab is poured onto reusable formwork on site in its final position. The formwork will be required to:

· be of the design shape,

· carry the construction load,

· be clean at the time of the concrete being poured,

· contain the concrete without leaking,

· strip easily.

To ensure the finished floor and construction tolerances are achieved, the formwork must accurately reflect the shape of the desired floor.

Formwork must remain in position until a significant amount of the specified strength has been achieved. The designer should advise a recommended stripping time.

It is extremely important to make the formwork sufficiently strong to be able to support the loads applied by the mass of the concrete, the reinforcement and working loads such as movement caused by people working on the slab and additional loads caused by concrete pumps or equipment.

★ Ⅲ. Group Activities

1. What steps should we follow to work on the concrete floors on fill?

2. Describe two methods used to support concrete floors poured between masonry walls.

扫一扫，听录音

The Pride of Chinese Architecture

Tian'anmen Square, once the main south gate of the imperial city during the Ming and Qing Dynasties. The Monument to the People's Heroes and the Chairman Mao Memorial Hall both lie on the axis of Tian'anmen Square, while the National Museum of China and the Great Hall of the People sit symmetrically on each side. This is an extension of traditional central axis planning. The square testi fies the great change from rule by the emperor to rule by the people and now becomes a major site for state activities.

Self-assessment

Now I know

· functional requirements of floor are________________________;

· the two types of floors are ____________________________;

· the two types of concrete floor systems poured on filling are __________;

· all concrete floors on hill should be reinforced with fabric______________;

· the floating slabs are used where__________________________;

· the qualities of insertion of a vapour barrier are____________________

____;

Now I can

- ☐ use the correct terminology in the description of floor construction;
- ☐ write a site report according to the workplace conversation;
- ☐ explain the functional requirements of floor;
- ☐ distinguish solid floor and suspended floor ;
- ☐ tell the course of damp proof.

Unit Eight

Finishes

Learning Objectives

After learning this unit, you will be able to

- describe ceiling finishes;
- write a site report according to the workplace conversation;
- explain demountable partitions and suspended ceiling systems;
- discuss finishes and coating;
- use the correct terminology in the description of finishes;
- understand Chinese excellent architectural culture.

Part one

Section A Workplace Conversation

扫一扫，听录音

Professional Words and Expressions

rectify ['rektɪfaɪ]	*v.*	调整；矫正
ceiling fixer		顶棚安装工
ceiling grid		顶棚龙骨
hold-down clip		固定卡子
spirit level		水平尺，水平仪
step ladder		人字梯，登高梯
suspension hanger		悬挂吊架

Ceiling Finishes

Accuracy is the finishing tradesman's chief skill. Most work on a building site does not require perfect accuracy. When an error is made, the decision to rectify it often depends upon whether the mistakes will be noticed.

★ I. Listen to the conversation and fill in the blanks with what you hear.

(Sid—Clerk of Works; Bernard—Ceiling Fixer)

Sid: I say, Bernard. I don't like the look of this at all. You haven't got the 1)____________ very level, you know. You'll have to do something about it, or I'll be reporting it to Peter.

Bernard: Oh, I'm sorry. Sid, I'm afraid I hadn't noticed it was out. I didn't fix this bit, you see. Tim did most of it and I haven't got around to checking it with the 2) ____________ yet.

Sid: You can see it's out quite clearly from up here on the 3) ____________. You'd better have a look yourself.

Bernard: Hmm, yes, it doesn't look right, I'll 4) ____________.

Sid: Will it be a lot of work to put it right?

Bernard: Well, that depends really. The 5) ____________ are adjustable, but obviously we have to take off the panels and 6) ____________ to get at them. It just depends how many need adjusting—it could take ages.

Sid: Oh dear.

Bernard: Couldn't we just leave it as it is. Sid, I don't think anyone notices. It looks almost level to me.

Sid: Oh no, definitely not. It just won't do, I'm afraid. There are going to be wall lights in here, so any sagging in the ceiling will really 7) ____________.

Bernard: Oh well. I suppose we'd better get on with it right away.

Sid: Yes. And don't forget the 8) ____________ want to start on this ceiling as soon as you've finished.

★ II. Sid writes a site report according to the conversation above.

This morning, I found Bernard's work on the ceiling is not acceptable. ______

Section B Demountable Partitions & Suspended Ceilings

扫一扫，听录音

Demountable partitions and suspended ceilings are often inter-related. Both should be based on the same modular frame of reference as to position. The method of restraint at the top of the demountable partition can affect the detail and construction of the ceiling. In practice, the demountable partition is often installed after the construction of the suspended ceilings.

1. Partition

Partition is a form of construction used to subdivide the space within the building, which includes demountable partitions (relocatable partition which can be moved with little damage when a redivision of the enclosed space is required), movable partition (partition which can be easily and quickly moved aside and not bearing on the walls) and suspended ceiling (a ceiling hung at a distance from the floor or roof above and not bearing on the walls). The following will mainly discuss demountable partitions and suspended ceiling.

2. Demountable Partitions

The majority of demountable partitions systems consist of a framework and infill panels.

The framework is made from either pressed steel, extruded aluminum, or hard / soft-wood. The Frame may be covered by the panel, exposed, or covered by a strip. Hollow metal frame may also act as ducting for services such as electricity and telephone.

Panels (including solid panels and glazed panels) are fabricated and manufactured off site. Solid panels may comprise of two sheets of plaster board or sandwich panels with the core materials of woodwool, expanded polystyrene, and eggcrate cardboard. Glazed panels are for view.

Demountable partitions provide buildings with much flexibility (in which a similarity of dimension maximizes the number of alternative positioning of any components), variable choice of appearance (in which the use of a related system of components greatly enhance the visual harmony of the whole). Demountable partitions are high performance interior partitions. And most of all, resulting from the use of standardized components and reduced site labour time (e.g. dry construction), they are cost saving.

Although the advantages of demountable partitions are significant, some of the elements must also be considered in the design.

1) fire resistance: the higher the demountability, the poorer the fire resistance.

2) Sound insulation: lightweight partition cannot be expected to have good sound insulation. Joints and gaps plus noise transmission through ceiling void will also affect performance. Therefore openings for doors, ventilation, services need to be carefully detailed.

3) Maintenance: this depends on choice of finishes (e.g. Laminated plastic sheet requires no maintenance while plasterboard requires painting).

3. Suspended Ceilings

A ceiling simply had been regarded as a single-plane, fire-protective, finished element overhead. However, with the introduction of a suspension system, the ceiling offered access to plumbing, electrical and mechanical components in overhead runs.

Today's suspended ceiling systems offer even more advantages for building construction, including a range of acoustical control options, fire protection, esthetic appearance, flexibility in lighting and HVAC (heating, ventilation, and air conditioning) delivery, budget control and optional use of overhead space.

A suspended ceiling system provides false ceiling hangings at a distance from soffit of the floor or roof above and not bearing on the wall, and it consists of a framework (void formed, usually of lightweight, steel or aluminum alloy section). A suspended ceiling system supports a variety of sheet panels and strips in such a system are fixed by dry construction method.

The performance of a suspended ceiling system requires the following elements, strength and stability (supporting fittings, fans, workers needed for services if required), access for maintenance (choice of ceiling type for access and special access door if needed), and acoustic control (sound absorption, sound reflections).

Suspended ceilings vary from simple grid systems to complex integrated ceilings. But basically suspended ceilings can be classified into the following categories.

1) Jointless suspended ceiling

· Plasterboard with skim coats.

· Expanded metal lathing soffit with hand applied plaster finish or sprayed applied rendering.

2) Paneled suspended ceilings

· These are the most popular form of a suspended grid framework to which the ceiling covering is attached.

· The covering can be of a tile, tray, board or strip format in a wide variety of material with an exposed or concealed supporting framework.

· Services such as luminaries can usually be incorporated within the system.

3) Strip ceiling

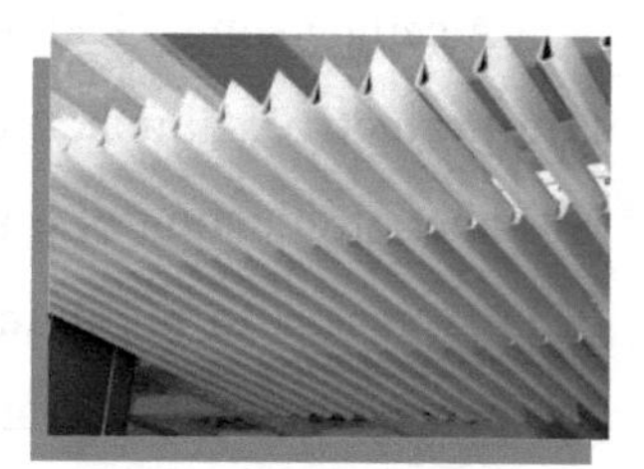

· Long narrow profiled aluminum tray or metal core with PVC facing.

· Can be fixed abut each other or interlocking along the edges or spaced apart.

· Fixed directly onto runners, do not require noggings.

4) Opening ceilings

· Bottom face of a pattern of baffles or eggcrate panels.

· Large range of materials; various patterns and finishes available.

· To allow concealed lighting to pass directly downwards through the ceiling.

5) Integrated suspended ceilings

· Ceiling systems specially designed to accommodate services with integral parts.

· Most designed to work with standard building planning grids of 1200mm or 1500mm.

· All are intended to be used in commercial applications, where the partitions, lights, and other elements connected with ceiling change frequently.

Professional Words and Expressions

demountable [di:'maʊntəbl]	*adj.*	可拆卸的，可卸下的
modular ['mɒdjələ(r)]	*adj.*	模数,（建筑等）组合式的，模块化的
restraint [rɪ'streɪnt]	*n.*	固定，约束，抑制
subdivide ['sʌbdɪvaɪd]	*v.*	再分，细分
strip [strɪp]	*n.*	长条，板条
ducting ['dʌktɪŋ]	*n.*	管道，导管
laminate ['læmɪnət]	*adj.*	由薄片叠成的
run [rʌn]	*n.*	管线布局
soffit ['sɒfɪt]	*n.*	拱腹；下表面；底面
alloy ['ælɔɪ]	*n.*	合金
rendering ['rendərɪŋ]	*n.*	罩面砂浆，初涂，打底
format ['fɔ:mæt]	*n.*	样式；规格
luminary ['lu:mɪnərɪ]	*n.*	照明灯，发光体
abut [ə'bʌt]	*n.*	平接，对头结合；支座，支架
runner ['rʌnə(r)]	*n.*	（吊顶）横龙骨，龙骨，中间支承干
nogging ['nɒgɪŋ]	*n.*	横撑杆件，木砖，木架砖壁（木架中填砖）
baffle ['bæfl]	*n.*	隔板，挡板；遮护物
woodwool ['wʊdwʊl]	*n.*	木丝，木刨花
polystyrene [ˌpɒli'staɪri:n]	*n.*	聚苯乙烯
eggcrate ['egkreɪt]	*n.*	蛋形格栅，花格
cardboard ['kɑ:dbɔ:d]	*n.*	硬纸板
infill panel		填充板
pressed steel		压制型钢，冲压型钢
extruded aluminum		挤压铝型材，压制铝材
glazed panel		玻璃板
false ceiling		假平顶，吊顶
skim coat		罩面层；薄覆盖层
profiled aluminum tray		预制铝合金槽

Exercises

★ Ⅰ. Answer the following questions.

1. What are the similarities between the demountable partitions and suspended ceilings?

2. What is the function of the partition?

3. What are the disadvantages of the demountable partitions?

4. What advantages do the suspended ceilings offer for building construction?

5. How is the suspended ceiling constructed?

★ Ⅱ. Match the following words or phrases with the correct Chinese.

1. partition	a. 板条
2. nogging	b. 隔墙
3. eggcrate	c. 填充板
4. strip	d. 木砖
5. false ceiling	e. 花格
6. runner	f. 拱腹
7. soffit	g. 预制铝合金槽
8. infill panel	h. 横龙骨
9. demountable	i. 可拆卸的
10. profiled aluminum tray	j. 吊顶

★ Ⅲ. Read the following pictures.

1. Find a proper word or phrase to describe each of the pictures.

1) 

2) ______________________________

3) 　　4) ______________________________

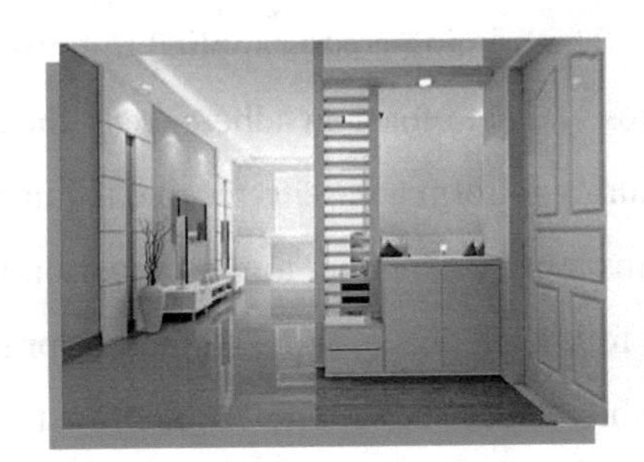

5) 　　6) ______________________________

2. Read the pictures again and tell your partners which ones are related to each other and what kind of construction work they show.

Part Two

Finishes and Coatings

The type of paint and the appropriate painting system depends on the following factors:

· the nature of the surface to be painted,

· the type of environment the finished paint will be subjected to,

· the cost,

· if the finish has to be simply decorative or hard wearing and washable.

Different paint types have specific roles to play in a painting system, and it is necessary for you to understand and appreciate their importance.

Sealers

Sealers are generally applied in painting to prevent undue absorption of the binder from subsequent coats of paint. They also help to prevent chemicals in the substrate material causing premature failing of the applied paint and to prevent previous paint from bleeding into subsequent coats of paint.

Primers

Primers are applied to help good adhesion of the paint system to the substrate. Different primers perform under different conditions of service. For example, wallboard primers are different to metal primers and metal etch primers and are required to have different performance characteristics.

The priming coat is applied to timber to reduce its porosity and to form an adherence to the surface, which in turn will form a base to receive the undercoat. It also forms a protective coat to the timber until such time as the following coats can be applied. A priming coat also inhibits things such as the action of rust on iron.

It will be apparent that the priming coat is the foundation of the job and upon it depends the quality and adhesive power of the coats to follow. Because this coat is covered many people think that it is an opportunity to get rid of all the odds and ends of pain that have been left over from previous jobs.

Undercoats

Undercoats are applied as an intermediate coat between a priming coat or a primer/sealer coat and the top or finish coat. The function of an undercoat is to provide a well filled surface on which to apply the topcoat, especially where timber is being painted. Undercoats also act as a sealer over any putty, stopping or fillings and will allow the finishing coat to dry without soakage blemishes.

Within limits, as the thickness of the paint increases so does its protective ability, therefore one of the functions of the undercoat is to impart thickness or body to the paint system. The face of timber in pored woods is pitted with minute cavities and the application of the priming coat is insufficient to fill these cavities. The undercoat therefore assists in filling and leveling the surface and also in obliterating any markings that would mar the finished appearance of the job.

Priming coats are usually pink in colour, so the undercoat needs to be closer to the colour of the finishing coat to cover this and to be capable of being covered by the finishing coats. The undercoat should not adhere to the primer, neither should it be so soft that the finishing coat will sink into it, causing a patchy appearance.

Finishing Coats

Finishing coats need to form a protective coat according to the conditions that they will be subjected to. No paint has yet been produced that will stand up to all the conditions that may be imposed upon it. Some of the conditions encountered will be rain, sun, sea air, frost, wind driven dust, chemical fumes, abrasion caused by washing, mould growth and moisture penetration. However, most paints stand up to these destructive influences for a number of years.

Another important function of the finishing coat is to impart the right texture and colour to the surface and to further protect the structure itself against deterioration.

Although not always possible, painting should be carried out during the late spring. The hot summer sun is likely to blister the paint, and frosts are likely to damage it. The timber is fairly dry during late spring and water will not be locked in the timber when painted over.

Professional Words and Expressions

finish ['fɪnɪʃ]	*n.*	外表装饰，表层，面层抹灰
coating ['kəʊtɪ]	*n.*	表面（保护或抹光）层
paint [peɪnt]	*n.*	油漆；涂料
sealer ['si:lə(r)]	*n.*	密封剂，封底漆
binder ['baɪndə(r)]	*n.*	基料；粘合剂
substrate ['sʌbstreɪt]	*n.*	基层，基质
primer ['praɪmə(r)]	*n.*	（用于基层保护或与底漆结合的）打底漆
adhesion [əd'hi:ʒn]	*n.*	粘连；粘合
porosity [pɔ:'rɒsətɪ]	*n.*	孔隙度；孔隙率
adherence [əd'hɪərəns]	*n.*	粘合；粘连
surface ['sɜ:fɪs]	*n.*	表面
undercoat ['ʌndəkəʊt]	*n.*	面漆下涂层；底涂层
inhibit [ɪn'hɪbɪt]	*v.*	禁止，抑制
rust [rʌst]	*v.*	生锈
putty ['pʌtɪ]	*n.*	腻子；堵料
soakage ['səʊkɪdʒ]	*n.*	浸润，浸透，浸渍

blemish ['blemɪʃ]	*n.*	瑕疵，缺点
thickness ['θɪknəs]	*n.*	厚度
level ['levl]	*n.*	水平线，水平面
obliterate [ə'blɪtəˌreɪt]	*v.*	除去；除掉
mar [mɑː(r)]	*v.*	毁坏
patchy ['pætʃɪ]	*adj.*	不调和的；拼凑成的
abrasion [ə'breɪʒn]	*n.*	磨损；磨耗
deterioration [dɪˌtɪərɪə'reɪʃn]	*n.*	衰退
premature failing		过早损坏
wallboard primer		墙板底漆
metal etch primer		金属磷化底漆
metal primer		金属底漆
protective coat		保护涂层
intermediate coat		二道底漆，中层漆
minute cavity		细微孔洞，微小空洞
mould growth		长霉
moisture penetration		潮湿的侵袭

Exercise

★ Ⅰ. Reading Comprehension

1. What does the type of paint and the appropriate painting system depend on?
2. In what order do we usually do the finish and coating?
3. When will primers be put into use? What are the functions of primers?
4. How can we make good undercoats?
5. What main factors affect the quality of the finishing coats?

★ Ⅱ. Translation

A. Translate the following sentences into English.

1. 不同的涂料类型有不同的功效，因此了解它们的特性很重要。
2. 涂料中加入密封剂，是为了防止涂料过度地吸收后续涂料中粘合剂。
3. 用于木方中的底漆是为了减小其孔隙并在表面形成粘贴。
4. 显而易见，底漆是粉刷墙面的基础，它决定着墙面的质量和涂料粘附力。

5. 炎热的夏天很可能会使油漆起泡，霜冻也有可能损坏油漆。因此，油漆工程应该尽可能在晚春进行。

B. Translate the following paragraph(s) into Chinese.

Finishing coats need to form a protective coat according to the conditions that they will be subjected to. No paint has yet been produced that will stand up to all the conditions that may be imposed upon it. Some of the conditions encountered will be rain, sun, sea air, frost, wind driven dust, chemical fumes, abrasion caused by washing, mould growth and moisture penetration. However, most paints stand up to these destructive influences for a number of years.

★ Ⅲ. Group Activities

1. Why is it important to know the differences between all these materials discussed in the passage above?

2. What are the most important factors in deciding what materials should be applied?

扫一扫，听录音

The Pride of Chinese Architecture

The Water Cube: The National Aquatics Center, also known as "the Water Cube" and "Ice Cube", is located in Beijing Olympic Park. It is a boutique venue for the 2008 Beijing Olympic Games and a classic transformed venue for the 2022 Beijing Winter Olympic Games.

The building's structural design is based on the natural formation of soap bubbles. To bring the design to life, the individual bubbles are incorporated into a plastic film and tailored like a sewing pattern. An entire section is pieced together and then put into place within the structure. There are interior and exterior films, and the film is then in flated once it is in-situ. It will be continuously pumped thereafter.

The National Aquatics Center is an international advanced center integrating swimming, sports, fitness and leisure.

Self-assessment

Now I know

· partition usually includes ____;

· demountable partitions include ____;

· suspended ceiling includes ____;

· panels usually include ____.

· although not always possible, planting should be carried out during ____.

Now I can

☐ use the correct terminology in the description of finishes;

☐ write a site report according to the workplace conversation;

☐ describe ceiling finishes;

☐ distinguish finishes and coating;

☐ explain the function of partition.

Unit Nine

Building Services

Learning Objectives

After learning this unit, you will be able to

- describe the three kinds of building services;
- write a site report according to workplace conversation;
- discuss electrical and plumbing installation;
- know the importance of electrical and plumbing installation;
- use the correct terminology in the description of building services;
- understand Chinese excellent architectural culture.

Part One

Section A Workplace Conversation

扫一扫，听录音

Professional Words and Expressions

ductwork ['dʌktˌwɜ:k]	*n.*	管道系统，管道
air conditioning [eə(r) kən'diʃəniŋ]	*n.*	空气调节装置；冷气
constant flow rate controller valve		恒流量控制阀
duct-mounted silencer		管式消声器
fire damper		防火档板；防火阀；防火风门
flexible ducting		柔性管道
intumescent material		发泡型防火材料
honeycomb		蜂窝

Ventilation

A building is an assembly of parts. Installation of a complicated air conditioning system using many components is a job which needs to be well organized.

★ **I. Listen to the conversation and fill in the blanks with what you hear.**

(Terry —Ductwork Installer; Alan—Apprentice Ductwork Installer)

Alan: Oh, yes.

Terry: We've got the actual 1) ____________down here already, but all the smaller parts we'll be needing are still in the store.

Alan: Oh.

Terry: I'll just explain what all these are, shall I? Then I'd like you to go and fetch them, Okay?

Alan: Oh, right.

Terry: Now, look at that symbol there- that's a duct-mounted 2)____________. Do you know what it looks like? —it's sort of 3)____________- cylindrical thing.

Alan: Hmm.

Terry: This one here—this is a 4)____________made from intumescent material. It's flat and round. Have you ever seen one? It looks 5) ____________.

Alan: Oh, yes.

Terry: And that one is a length of flexible ducting, and there is a 6)____________.

Alan: Hmm. Right.

Terry: Are you listening, Alan? You don't seem to be paying much attention.

Alan: Sorry, Terry. Actually, er—I was just wondering—would you mind if I left a bit early this afternoon?

Terry: Well, there's a lot to do today. We've got to 7) ____________by the end of the afternoon, you know. Is it important?

Alan: Well—my brother's in hospital. I was hoping to go and see him straight from work.

Terry: Hmm. I wouldn't mind you're leaving early sometimes if you got here on time in the mornings. Look, I'll tell you what. Let's see how the work goes today. You'd better try and make a good start this morning. If you get on okay, I'll let you go early this afternoon.

Alan: Oh, thanks, Terry. Well, I'd better get started then. What should you like me to do first?

Terry: Let's see, now. You 8)________________________and make yourself a list of all these parts we need. And if there's anything you don't understand, just let me know.

★ Ⅱ. Terry writes a site report according to the conversation above.

This morning, I instructed Alan to install the air conditioning system. ________

____

____

____

____

____

____

____

____

Section B Building Services

扫一扫，听录音

Building services include environmental services, utility services and building services systems. Environmental services generally refer to thermal such as heating, air-conditioning and ventilation which provide thermal balance; lighting including both natural day lighting and artificial lighting; acoustics such as noise, room effects and sound reinforcement systems.

Utility services are generally referred to as follows:

1) water supply — both hot and cold;

2) electrical supply and telecommunications;

3) drainage — roof water plumbing and drainage;

4) sewage disposal — sewerage plumbing and drainage;

5) refuse collection and storage;

6) gas supply;

7) mechanical conveyors — lifts, escalators;

8) fire protection — alarms, sprinklers, fire escape.

Building services systems include:

1) building automatic system;

2) building security system;

3) intelligent building system.

The connection of services to a modern dwelling is an extremely important part of a building contract. Most of the services are controlled by specific statutory bodies and require installation to be carried out by a licensed tradesperson.

The two trades you are most likely to deal with on a construction site are the plumber that will be mentioned in part two and electrician as follows.

All electrical installation in the construction of a building must be carried out by a qualified licensed electrician. If the installation of power to a building is incorrectly carried out there is the possibility of both electrocution and fire caused by an electric short.

The following description covers the most important terms and shows where they are situated in the installation.

The mains (main): This is the supply from the supply authority. The mains can be overhead on a power pole or underground and the connection made in a service pit. They are usually low voltage—415/240 volts—but some large users of electricity have a high voltage mains.

The service: This is connected from the mains to the point where the customer's wiring (service cable) starts (consumer's terminals).

Service pit: This is an underground area only. It is the cylindrical pit buried near the boundary of two properties that has the supply authority mains and the connection to the customer's wiring (the consumer's mains). There is one pit for every two customers.

Service fuse: This is the fuse, supplied by the supply authority, which is used to isolate the installation from the mains if there is a large fault or if there is a need to disconnect the installation for any reason. Normally the customer's wiring is connected to one side of the fuse and the supply authority's service is connected to the other.

Consumer's terminals: This is the junction between the supply authority's wiring with the customer's wiring. The consumer's terminals are usually in the service fuse or where the connection to the mains is made in service pit. The supply authority is responsible for the wiring up to the consumer's terminals. From then on it is the responsibility of the customer.

Consumer's mains: This is the cable which runs between the consumer's terminals and the main switchboard. It is the cables that run from their service fuse on the fascia, through the ceiling, to the meter box and on to the switchboard. Or in an underground area it is the cables that run from the service pit under the front yard to the meter box and on to the switchboard.

Meter enclosure: This often referred to as the meter box. It houses the metering instruments and any other devices provided by the supply authority. Sometimes for convenience the main switch board is included in the meter enclosure.

Switchboard: The switchboard has the circuit protection devices—fuses or circuit breakers—as well as the main earth connection. If there are a number of switchboards in the installation, such as when there is a shed or an outbuilding of some kind, the main switchboard is the one which is supplied by the consumer's mains and it also has the main earth connection. The other switchboards are called sub-boards or distribution boards.

Circuit: This is one run from the switchboard which feeds other boards, outputs or appliances. Each circuit has its own protection device, so if there are five fuses on the switchboard, then there are five circuits coming from that switchboard.

Sub-circuit: This is the feed to a sub-switchboard, the sub-switchboard will feed other parts of the installation, usually separate from the main area.

Final sub-circuit: The supply to the outlets—power points, lights, or fixed appliances such as the hot water service – is called the final sub-circuit. There are no other circuits originating from this circuit.

GPO: GPO is short for a General Purpose Outlet. This is a socket outlet which is used for general use, such as in the kitchen, the bedroom etc. it is not an outlet which has a specific use, such as for an air conditioner. These are called SPOs, Special Purpose Outlets.

Fixed appliance: A fixed appliance is one that is fastened to a support or secured into position, such as the hot water service, an air conditioner, the stove and some heaters. The supply to these is usually connected through a connecting terminal block rather than with a plug and socket.

Earth: This term is usually used with other terms such as main earth or earthing conductor or the earthing system. As the supply authority uses the earth as a conductor, for safety every conductive part of the installation, other than the live conductors, is connected to the earth (or ground). This is done with earth wires which are connected by various means to the ground, such as with an earthing stake. So the earth refers to the conductive ground and the earth wire is the wire that connects to the ground form parts that require earthing, like the metal surround of the stove, or the earth pin of a GPO.

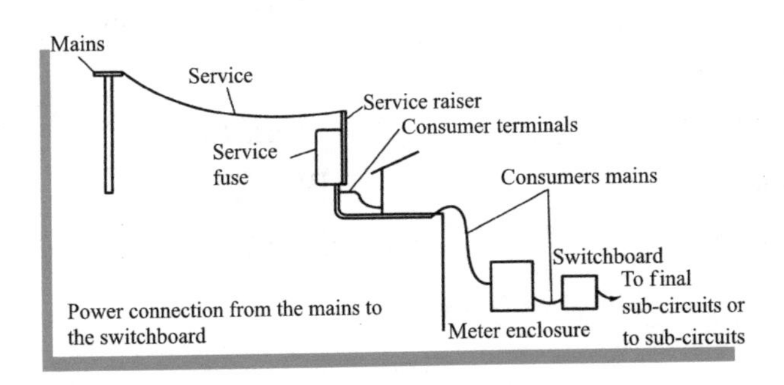

Power connection from the mains to the switchboard

Professional Words and Expressions

thermal ['θɜ:məl]	*n.*	冷热，温度
ventilation [ˌventɪ'leɪʃn]	*n.*	通风设备；空气流通
acoustics [ə'ku:stɪks]	*n.*	声学；（传声系统的）音响效果
refuse [rɪ'fju:z]	*n.*	垃圾，废弃物
conveyor [kən'veɪə(r)]	*n.*	输送设备；传送带
escalator ['eskəleɪtə(r)]	*n.*	自动扶梯
sprinkler ['sprɪŋklə(r)]	*n.*	（建筑物内的）自动喷水灭火装置
statutory ['stætjʊtrɪ]	*adj.*	法定的，法令的
electrocution [ɪˌlektrə'kju:ʃn]	*n.*	触电，触电死亡
voltage ['vəʊltɪdʒ]	*n.*	电压，伏特数
cylindrical [sə'lɪndrɪkl]	*adj.*	圆柱形的，圆筒状的
fuse [fju:z]	*n.*	保险丝；导火线
switchboard ['swɪtʃbɔ:d]	*n.*	配电盘，控制板
circuit ['sɜ:kɪt]	*n.*	电路，线路
socket ['sɒkɪt]	*n.*	插座；灯座
utility service		公用服务设施
building service		建筑设备
sewage disposal		污水处理

supply authority mains	电源供应局
meter box	仪表箱，电表箱
circuit breaker	断路开关，断路器
earth connection	接地线
distribution board	配电板，配电盘
sub-circuit	子电路，支回路
General Purpose Outlet	通用插座
Special Purpose Outlet	专用插座
fixed appliance	固定设备
consumer terminal	用户终端

Exercises

★ Ⅰ. Answer the following questions.

1. What do building services generally refer to?
2. Who are entitled to provide building services?
3. How does the fuse work?
4. What do GPO and SPOs mean respectively?
5. What is the consumer's mains and how do the cables run through?

★ Ⅱ. Match the following words or phrases with the correct Chinese.

1. earth connection	a. 输电干线
2. mains	b. 固定设备
3. service fuse	c. 配电板，配电盘
4. distribution board	d. 仪表箱，电表箱
5. conductor	e. 客户终端
6. fixed appliance	f. 电源供应局
7. meter box	g. 接地线
8. consumer terminals	h. 入户保险丝
9. supply authority mains	i. 公用服务设施
10. utility services	j. 导体

★ Ⅲ. Read the following pictures.

1. Find a proper word or phrase to describe each of the pictures.

1) ______________________

2) ______________________

3) ______________________

4) ______________________

5) ______________________

6) ______________________

2. Read the pictures again and tell your partners which ones are related to each other and what kind of building services they show.

Part Two

扫一扫，听录音

Plumbing Fix

A plumber must be registered to install the sewerage system, hot and cold reticulated water systems and gas services. One single tradesperson may not do all this work. For example, a registered drainer may carry out the installation of the sewerage system while a registered plumber will direct the installation of the hot and cold water systems.

Sewerage Plumbing and Drainage

The sewerage plumbing and drainage is carried out by a sanitary plumber or drainer. They are responsible for the installation of the sewerage system or septic tank in a non-sewerage area. Sewerage plumbing and drainage must be carried out by a licensed plumber in order to ensure that standards are maintained.

The purpose of a sewerage system is to remove septic and effluent wastes from a building. This removal is provided through a collection of pipes and assorted fittings that combine to remove the liquid waste.

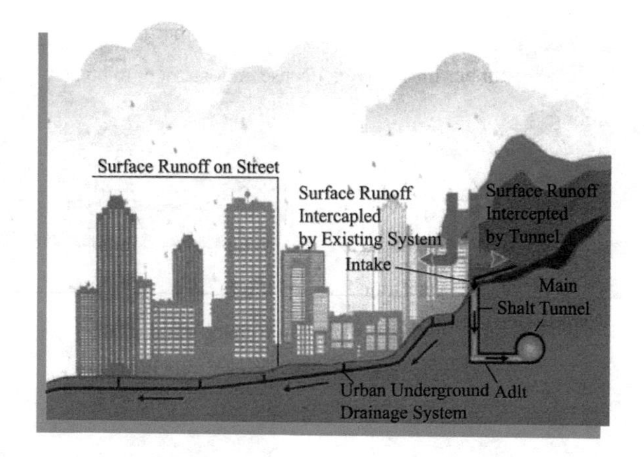

Plumbing First Fix

The plumber is also responsible for the installation of the cold and hot water reticulation within the house and the gas service where applicable.

Almost without exception, where piped town water is supplied to a property, there is a requirement that the water reticulation system be installed by a licensed plumber.

Cold Water Reticulation

When using town supply the first task of the plumber is to bring water from the water supply authority's main to the property water meter. The water meter is generally at or near the front boundary of the property. From the water meter the water supply is connected to the cold water points in the dwelling.

The pipe used for cold water reticulation is generally copper or stainless steel, but in recent times some water supply authorities have approved the use of special plastic pipes and fittings.

The plumber may either place the cold water pipes under the house, in the walls or in the roof space, whichever is the most convenient. The plumber must take special care to ensure the pipes are saddled or fixed adequately to the building.

Hot Water Reticulation

As with cold water reticulation it is normal for the hot water to be piped in copper or stainless steel pipes, the only difference being that the hot water supply pipes are insulated. Traditional hemp type fibre insulations are still used by some plumbers, but modern synthetic foam insulation is more efficient.

Hot Water Units

Water heaters are generally classified according to their working pressure as:

· mains

· reduced

Mains pressure units are designed for direct connection to cold water mains so that hot water is delivered at the same pressure as cold water.

Reduced pressure units are designed so that the pressure is reduced by a pressure limiting valve, reduction valve or by overhead feed tanks. A variation to this is the low pressure system in which pressure in the system is maintained at atmospheric pressure by means of an overhead mounted feed tank attached to the system.

Gas Fitter

The plumber is also responsible for the installation of the gas service from the gas meter to the building. A plumber must have a license endorsement to be able to install gas services to a building.

Plumbing Second Fix

w fix involves the connection of sanitary fittings to the sewerage system, and is considered to start at the floor level of the building. The plumber assembles numerous precision-made rigid plastic pipes, traps and junctions together by way of threaded joints, patented connectors or simple glue.

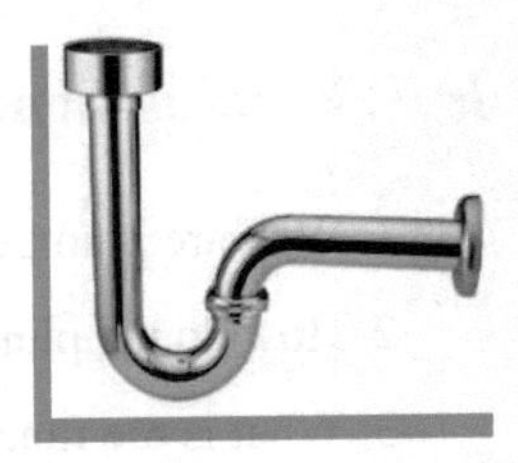

Professional Words and Expressions

meter ['mi:tə(r)]	*n.*	表；计量器
trap [træp]	*n.*	存水弯，水封
junction ['dʒʌŋkʃn]	*n.*	接合点，连接点

plumbing fix	管道安装
sewerage system	污水系统，污水排放系统
hot and cold reticulated water system	冷热水循环系统
gas service	燃气设施
sewerage plumbing	污水管道工程，下水道系统
septic tank	化粪池
non-sewerage area	非排污水区
licensed plumber	执照管道工，注册管道工
septic and effluent waste	污水废水
assorted fitting	配套管件，组合管件
water reticulation system	水网系统
hemp type fibre insulation	麻制类纤维绝缘材料
synthetic foam insulation	合成泡沫绝缘材料
mains pressure	总管压力
reduced pressure	减压
pressure limiting valve	限压阀
feed tank	给水箱
second fix	二次安装
precision-made rigid plastic pipe	精密制造的硬质塑料管
threaded joint	丝扣接头；螺纹接头
patented connector	专利连接件，特制接头

Exercises

★ Ⅰ. Reading Comprehension

1. Why are plumbers important in the installation of building service systems?
2. How do the plumber place pipes for cold water reticulation?
3. What is the function of the mains pressure units? How does it work?
4. What should a plumber know about hot water units?
5. By what way can a plumber assemble the pipe systems?

★ Ⅱ. Translation

A. Translate the following sentences into English.

1. 排水系统，冷热水循环系统和燃气设施必须由注册水管工安装。

2. 最近，有些供水部门已经批准冷水管道可以使用特殊的塑料管道或管件。

3. 污水处理系统的作用是排除建筑物中的污水废水。

4. 虽然传统的麻制类纤维绝缘材料仍用于管道中，但是现代的合成泡沫绝缘材料更便捷有效。

5. 要使用城市自来水，管道工的首要任务是将水从供应商的总表接到用户水表上。

B. Translate the following paragraph(s) into Chinese.

When using town supply the first task of the plumber is to bring water from the water supply authority's main to the property water meter. The water meter is generally at or near the front boundary of the property. From the water meter the water supply is connected to the cold water points in the dwelling.

The pipe used for cold water reticulation is generally copper or stainless steel, but in recent times some water supply authorities have approved the use of special plastic pipes and fittings.

The plumber may either place the cold water pipes under the house, in the walls or in the roof space, whichever is the most convenient. The plumber must take special care to ensure the pipes are saddled or fixed adequately to the building. A banging water pipe (water hammer) attached to the timber framework in the cavity of a brick veneer wall can be very expensive to correct.

★ Ⅲ. Group Activities

1. What should be born in mind when plumbing fix is being conducted?

2. What are the differences between cold water reticulation and hot water reticulation?

The Pride of Chinese Architecture

扫一扫，听录音

Hong Kong-Zhuhai-Macau Bridge (HZMB) is both the longest sea crossing and the longest fixed link on earth. It comprises a series of bridges and tunnels that connect Hong Kong, Macau, and Zhuhai in the Pearl River Delta in China. The overall length is about 55 km (34 mi), with the main section including a 29.6 km dual three-lane carriageway in the form of a bridge-cum-tunnel structure and two man-made islands. The bridge reduces the transit time from Hong Kong to Zhuhai from several hours to around 30 minutes. With great difficulty in construction, it is known as "Mount

Everest" in the bridge industry, and is also rated as one of the "new seven wonders of the world" by the British guardian.

Self-assessment

Now I know

· building services include ____;

· utility services include ____;

· building services systems include ____;

· plumbing installation include ________________________;

· the most important terms in electrical installation include ____.

Now I can

☐ use the correct terminology in the description of building services;

☐ write a site report according to the workplace conversation;

☐ describe the three kinds of building services;

☐ distinguish the constituent parts of a building;

☐ explain electrical and plumbing installation.

Unit Ten

External Works and Landscaping

Learning Objectives

After learning this unit, you will be able to

- describe external works;
- write a site report according to the workplace conversation;
- explain landscaping gardening;
- have basic ideas of surface drainage;
- use the correct terminology in the description of external works and landscaping ;
- understand Chinese excellent architectural culture.

Part One

Section A　Workplace Conversation

扫一扫，听录音

Professional Words and Expressions

handover ['hændəʊvə(r)]	*n.*	移交
pave [peɪv]	*v.*	铺设
gang [gæŋ]	*n.*	一组（人）
hold-up [həʊld ʌp]	*n.*	耽误
fall [fɔ:l]	*n.*	坡度，斜度，坡降
ramp [ræmp]	*n.*	坡道
external works		室外工程，外部工程
layout drawing		平面布置图，规划图
drainage channel		排水沟

The Last Minute Rush

It is often necessary to leave all the external works on site till last, so that they will not be damaged while other work is being completed. Sometimes the last jobs have to be hurried in order to be finished on time.

★ Ⅰ. Listen to the conversation and fill in the blanks with what you hear.

(Stanley—Foreman Pavior ; Paul—Pavior)

Stan: Morning, Paul. Do you know Peter has just told me the 1)________ for the offices has been brought forward to the end of next week?

Paul: The end of next week? — You are joking. We've got to 2)__________ the whole of the courtyard first, haven't we?

Stan: Yes, of course we have. That's just what I came to see you about. Er—I wondered if you'd mind working evenings next week.

Paul: Well, I'm not very 3)__________ working overtime just now, Stan. I told you Sandra started working nights at the hospital, didn't I? So we don't see that much of each other at the moment. She gets really 4)_________ with being on her own in the evenings.

Stan: Oh, go on, Paul. It's only a week, you know. There's no point getting another gang here just for a day or two. It'll be good money for you, too.

Paul: Well. I'm sorry, Stan, but I would prefer not to if you don't mind.

Stan: But look here, Paul. 5) __________________

Paul: Yes, but a week early? It's too much to ask.

Stan: But you know how hard everyone else has had to work to keep on programme.

Paul: Hmm.

Stan: And we've got a really good set of drawings this time. There won't be any 6)__________ on this job, I promise you.

Paul: Well, I don't care. I ...

Stan: Look, Paul, we've got full 7)__________ here showing where everything is—the falls, the drainage channels, everything. And look at these sections showing all the 8)__________ of the ramps, all the edge details—we've got all we need to know.

Paul: Yes, but I ...

Stan: And Lane says all the other gangs are busy at the moment, you see. So I told him I thought you'd help out. Please, Paul.

Paul: Oh all right, Stan. I'll do it. But I don't know what Sandra's going to say. And I'll tell you this—it'll be the last time you'll get me to do anything I don't want to.

Stan: Thanks, Paul.

★ Ⅱ. Stan writes a site report according to the conversation above.

This morning I went to the working site to have a talk with Paul for the changing of handover time.____

____

____

____

____

Section B External Works

扫一扫，听录音

External works can cover a broad range of items and will generally include the following:

- paving to driveways and paths,
- disposal of surface water,
- retaining walls where the site is cut-and- fill,
- landscaping gardening.

Paving

Paving is not only a way of decorating an area around a house. It also provides a solid surface for areas such as garden paths, driveways, barbecue areas, patios and the like.

The main considerations when choosing paving materials are:

- the function it is to perform,
- its safety under the conditions of use,
- the site conditions,
- the aesthetic qualities desired,
- the initial cost,
- the cost of maintenance.

Where recommended by the footing designer, paving should be laid around the perimeter of the house. It should be a recommended width or a minimum of 600 mm and graded to drain any surface storm water away from the building without pooling.

There are five common types of paving materials which include:

- gravel,

· concrete,

· brick,

· concrete pavers,

· stone.

Gravel is one of the cheaper forms of paving. It is hardy and provides an excellent surface for driveways and large areas where there is no need to go to great expense. It is available in a number of colours and textures and is easy to install.

Concrete is a hard surface that is excellent for driveways, paths and other large areas. However, it is simple in its appearance and generally requires the services of an experienced tradesman for an ornamental finish.

Bricks are commonly used to obtain a quality paved finish in areas designed to complement the look of a building. They come in a variety of colours and styles and can be laid in ways limited only by your imagination. Paving with bricks can be time consuming and it is important that an even surface be prepared before the pavers are laid. The ground preparation can be the most important stage of paving.

Depending on the type of stone chosen, the final look can be either ornamental or rugged. Stone can be quite expensive but there is also imitation stone available for those who like the look but not the cost.

Disposal of Surface Water

Rainwater must be dispersed in a way to prevent ponding and water damage occurring to a building, its footings and surrounding structures. Subsoil drains may be required on the uphill side of cut and fill sites, adjacent to deep strip footings, behind retaining walls, adjacent to basement walls etc. in soil with poor drainage qualities or excessive subwater movement.

It is important that any stormwater run-off does not:

· cause water to enter the building,

· affect the stability of the building or any surrounding building,

· cause dangerous conditions on the site.

Things such as building, paving, and excavation all affect the natural drainage of rainwater and the installation of an alternative drainage system may be necessary.

Retaining Walls

Retaining walls are built to restrain a mass of earth or other materials. They must be safe against overturning and sliding forward, and the pressure under the toe (front bottom edge of footing) should not exceed the bearing capacity of the soil which it rests upon. In addition, the friction between the footing and soil-plus the pressure of any earth in front of the wall must be sufficient to keep the retaining wall from sliding forward. Finally, the wall must be strong enough to prevent failure at any point in its length due to the force of the retaining material.

Landscaping Gardening

Landscaping generally includes hard landscape and soft landscape. Hard landscape is such as building driveway, road, footpath, access, patio, fence, drainage channel, pond, or services of lighting column, litter bin, seat, sign, etc. Soft landscape is such as planting trees, grasses, flowers, plants, container-grown plants, etc.

Weather conditions are important on a building site because they have a great effect on whether work can go on as planned. Usually it is impossible to carry on working without cover in high winds, sub-zero temperatures, heavy rainfall and extremely hot weather. Sowing seed and planting trees and shrubs can only be carried out at the right time of the year. Landscape work is therefore difficult to fit into a building programme.

Professional Words and Expressions

aesthetic [i:s'θetɪk]	*adj.*	审美的
drain [dreɪn]	*v.*	排水
gravel ['grævl]	*n.*	砾石，碎石
rugged ['rʌgɪd]	*adj.*	粗糙的
disperse [dɪ'spɜ:s]	*v.*	（使）分散
cut-and-fill		挖填方
concrete paver		混凝土地面砖；混凝土铺路机
ornamental finish		装饰精修
imitation stone		人造石
subsoil drain		地下排水管（道），地下排水沟
landscaping gardening		园林园艺
hard landscape		硬绿化，硬质景观

soft landscape	软绿化，软质景观
lighting column	灯柱
litter bin	废物箱，垃圾箱

Exercises

★ Ⅰ. Answer the following questions.

1. What do external works generally include?
2. What main factors shall we consider when choosing paving materials?
3. What are the advantages of the gravel?
4. What is the function of retaining walls?
5. Why weather conditions are important on a building site?

★ Ⅱ. Match the following words or phrases with the correct Chinese.

1. gravel	a. 混凝土地面砖
2. concrete paver	b. 地下室
3. basement	c.（使）分散
4. retaining wall	d. 挖掘
5. disperse	e. 挡土墙
6. excavation	f. 挖填方
7. subsoil drain	g. 废物箱
8. litter bin	h. 地下排水沟
9. ponding	i. 砾石
10. cut-and- fill	j. 积水

★ Ⅲ. Read the following pictures.

1. Find a proper word or phrase to describe each of the pictures.

1) ______________________________

2) ______________________________

3) ______________________________

4) ______________________________

5) ______________________________

6) ______________________________

2. Read the pictures again and tell your partners what you know for each picture.

扫一扫，听录音

Part Two

Surface Drainage

Water resulting from rain which is collected or concentrated by either a building or site works must be disposed of in a way that avoids the likelihood of damage or nuisance to the building and any other surrounding property.

Stormwater should be diverted away from a building and never be allowed to discharge onto an adjoining property. Stormwater drainages to the street water table should conform with the local council requirements.

On a cut-and-fill site the surface water must be diverted away from any building with a slab on ground floor system by grading the finished surface away from the building. The ground around the building should be graded to give a slope of 50 mm over the first metre and finished both during the construction of the work, and at the completion, so that stormwater does not pond against or near the footings. It is important at all times during the construction of the building to make adequate provisions for the protection of the footings from stormwater.

Many site investigation and footing construction reports will specify this as a major consideration. For example:

Site Drainage

This building area is to be graded away from the house so that water drains away. On cut sites, ensure water can drain away along the base of the cut, provide a spoon drain if necessary. During construction provide temporary stormwater drains. Refer to general notes for additional information on this matter, as approved on Site Drainage Plan.

The minimum heights above the finished ground level of slab on ground floor systems are:

· 150 mm above natural ground level,

· 100 mm above sandy, well drained areas,

· 50 mm above paved or concrete areas that slop away from the building.

The slab heights above finished ground/paving level may vary depending on:

· local plumbing requirements. This will possibly relate to the height of the overflow relief gully relative to the drainage fittings and ground. To work effectively they must be a minimum of 150 mm below the finished floor slab level.

· the run-off from storms and the local topography.

· the effect of the excavation on a cut-and- fill site.

· the possibility of flooding.

· termite barrier provisions.

The ground beneath suspended floors must be graded so that the area beneath the building is above the external finished ground level and water is prevented from flowing under the building.

Professional Words and Expressions

nuisance ['nju:sns]	*n.*	损害
adjoining [ə'dʒɔɪnɪŋ]	*adj*	毗连的

specify ['spesɪfaɪ]	*vt.*	说明，明确提出
slope [sləʊp]	*n.*	斜坡；斜面
surface drainage		地面排水
ground floor system		底层结构
site investigation		现场调查，场地勘察
spoon drain		V形排水沟；匙状排水沟
site drainage plan		施工现场排水平面图
ground level		地面标高
paving level		路面标高
over flow relief gully		溢流排水沟
termite barrier provision		白蚁阻隔层规定

Exercises

★ I. Reading Comprehension

1. Why should the rain water be taken good care of on the work site of a building?

2. How should the surface water be treated on a cut-and- fill site?

3. What should be specified in a site investigation and footing construction report?

4. What are the minimum heights about the finished ground level of slab on ground floor system?

5. Why must the ground beneath suspended floors be graded?

★ II. Translation

A. Translate the following sentences into English.

1. 在建筑施工中，始终要有充分措施保护基础不受雨水的损坏，这是非常重要的。

2. 在开挖现场，为确保水沿着挖方底部排出，如有必要需提供一个匙状排水沟。

3. 悬垂楼板地基面必须有一定的坡度，这样建筑物地面就会高出外部装修后的路面，阻止雨水在建筑物下面流淌。

4. 排入城市街道地下的雨水水位应符合当地政府的规定。

5. 雨水应该疏离建筑物，但绝对不允许排放到毗邻的建筑物。

B. Translate the following paragraph(s) into Chinese.

On a cut-and- fill site the surface water must be diverted away from any building with a slab on ground floor system by grading the finished surface away from the building. The ground around the building should be graded to give a slope of 50 mm over the first metre and finished both during the construction of the work, and at the completion, so that stormwater does not pond against or near the footings. It is important at all times during the construction of the building to make adequate provisions for the protection of the footings from stormwater.

★ Ⅲ. Group Activities

1. In what way is drainage important on a working site?
2. What can we actually know from a site drainage plan?

The Pride of Chinese Architecture

扫一扫，听录音

Suzhou Gardens: Suzhou classical gardens which absorb the essence of Jiangnan garden architecture art are excellent cultural heritage of China, and are listed as human and natural cultural heritage by the United Nations. They are good at skillfully composing the limited space into changeable scenery, and win with small and exquisite in structure. They represent the style and artistic level of Chinese private gardens and are rare tourist attractions. They are integrated with houses and gardens, which can be enjoyed, visited and lived, and can experience people's comfortable life. The formation of this architectural form is a creation for human beings to cling to nature, pursue harmony with nature, beautify and improve their living environment in densely populated cities and cities lacking natural scenery.

Self-assessment

Now I know

· external works generally include ____;

· the common types of paving materials include ______________________________________;

· landscaping can be generally divided into ______________ and ______________;

· it is often necessary to leave all the external works on site till last, so that ____;

· the slab heights above finished ground/paving level may vary depending on ____.

Now I can

☐ use the correct terminology in the description of external works and landscaping ;

☐ describe external works;

☐ write a site report according to the workplace conversation;

☐ explain landscaping gardening;

☐ discuss surface drainage in group activities.

Appendix

Professional Words and Expressions

A

abrasion [ə'breɪʒn]	*n.*	磨损；磨耗	Unit 8
absorber [əb'sɔ:bə]	*n.*	吸收器，吸收装置	Unit 6
abut [əb'bʌt]	*n.*	平接，对头结合；支座，支架	Unit 8
access to the site		进入现场	Unit 1
accessible [ək'sesəbl]	*adj.*	可接受的	Unit 2
acoustics [ə'ku:stɪks]	*n.*	声学；（传声系统的）音响效果	Unit 9
adherence [əd'hɪərəns]	*n.*	粘合；粘连	Unit 8
adhesion [əd'hi:ʒn]	*n.*	粘连；粘合	Unit 8
adjacent [ə'dʒeɪsnt]	*adj.*	邻近的，毗邻的	Unit 4
adjoining [ə'dʒɔɪnɪŋ]	*adj*	毗连的	Unit 10
aesthetic [i:s'θetɪk]	*adj.*	审美的	Unit 10
air conditioning		空气调节装置；冷气	Unit 9
allotment [ə'lɒtmənt]	*n.*	供分配使用的土地	Unit 2
alloy ['ælɔɪ]	*n.*	合金	Unit 8
annotate ['ænəteɪt]	*v.*	注释，注解	Unit 1
appliance [ə'plaɪəns]	*n.*	设备；器具	Unit 4
architect ['ɑ:kɪtekt]	*n.*	建筑师	Unit 1
ash [æʃ]	*n.*	灰	Unit 7
asphalt ['æsfælt]	*n.*	（铺路等用的）沥青混合料	Unit 7
assembly [ə'semblɪ]	*n.*	装配，组装；装配图	Unit 1
assorted fitting		配套管件，组合管件	Unit 9

B

backfill ['bækfɪl]	*v.*	回填	Unit 3
baffle ['bæfl]	*n.*	隔板，挡板； 遮护物	Unit 8
balustrade [ˌbælə'streɪd]	*n.*	栏杆（柱），栏杆扶手	Unit 1
barrier ['bærɪə(r)]	*n.*	屏障，隔离，障碍	Unit 6
basement ['beɪsmənt]	*n.*	地下室，基础结构	Unit 3
batten ['bætn]	*n.*	板条；压条	Unit 4
beam [bi:m]	*n.*	梁，横梁，承重梁	Unit 4
bearing capacity		承重能力，承载力	Unit 3
bed joint		水平缝，层间接缝	Unit 6
bed [bed]	*v.*	铺砖，将……铺平	Unit 7

bedrock ['bedrɒk]	*n.*	基岩，岩床	Unit 3
binder ['baɪndə(r)]	*n.*	基料；粘合剂	Unit 8
biological decay		生物分解	Unit 7
birdsmouth [bɜ:dz'maʊð]	*v.*	开齿槽，角口承接，企口结合	Unit 5
bitumen [bɪ'tu:mən]	*n.*	沥青，柏油	Unit 7
blemish ['blemɪʃ]	*n.*	瑕疵，缺点	Unit 8
block [blɒk]	*n.*	砌块	Unit 1
bolt [bəʊlt]	*n.*	螺栓，螺钉	Unit 4
bolt [bəʊlt]	*v.*	拴住；螺栓	Unit 5
bond [bɒnd]	*v.*	砌合，结合，组合	Unit 6
bricklayer ['brɪkleɪə(r)]	*n.*	砖工	Unit 1
brickwork ['brɪkwɜ:k]	*n.*	砖砌体，砌砖，砌砖工程	Unit 6
buckle ['bʌkl]	*v.*	变形，弯曲	Unit 6
buckling		失稳	Unit 6
building service		建筑设备	Unit 9
bulge [bʌldʒ]	*v.*	凸出；膨胀	Unit 3
buyer ['baɪə(r)]	*n.*	采购员	Unit 1

C

caisson ['keɪsən]	*n.*	沉箱，沉井；	Unit 3
camber ['kæmbə(r)]	*n.*	拱形，弧形	Unit 7
capillary [kə'pɪləri]	*adj.*	毛发状的，毛细作用	Unit 7
cardboard ['kɑ:dbɔ:d]	*n.*	硬纸板	Unit 8
carpenter ['kɑ:pəntə(r)]	*n.*	木工	Unit 1
cast [kɑ:st]	*v.*	浇筑，铸造	Unit 4
cavity ['kævətɪ]	*n.*	（空）腔	Unit 3
cavity brick		空心砖	Unit 7
ceiling fixer		顶棚安装工	Unit 8
ceiling grid		顶棚龙骨	Unit 8
chiller ['tʃɪlə]	*n.*	制冷机组	Unit 5
chippie ['tʃɪpɪ]	*n.*	木工，木匠	Unit 4
circuit ['sɜ:kɪt]	*n.*	电路，线路	Unit 9
circuit breaker		断路开关，断路器	Unit 9
civil ['sɪvl]	*adj.*	土木的	Unit 1

cladding ['klædɪŋ]	*n.*	外墙板，外墙覆面板；围护结构	Unit 4
cleat [kli:t]	*n.*	托座；连接板件	Unit 4
clerk of works		现场监工员，工程代表	Unit 1
client ['klaɪənt]	*n.*	雇主	Unit 1
coating ['kəʊtɪ]	*n.*	表面（保护或抹光）层	Unit 8
collapse [kə'læps]	*n.*	崩塌，塌陷	Unit 3
column ['kɔləm]	*n.*	柱，立柱	Unit 3
commence [kə'mens]	*v*	开始	Unit 2
commencement [kə'mensmənt]	*n.*	开始	Unit 1
compact ['kɒmpækt]	*v.*	夯实，压紧，（使）坚实	Unit 2
compartment [kəm'pɑ:tmənt]		间隔，隔断；分隔间	Unit 6
component wall	*n.*	防火墙	Unit 6
component [kəm'pəʊnənt]	*n.*	零（部）件，构件	Unit 1
composite ['kɒmpəzɪt]	*n.*	合成物，混合物，复合材料	Unit 7
compression [kəm'preʃn]	*n.*	抗压；压迫	Unit 4
compressive force		压力	Unit 4
concave ['kɒn'keɪv]	*n.*	凹面，成凹形	Unit 4
conceal [kən'si:l]	*v.*	隐藏，遮住	Unit 7
concrete floor		混凝土底板	Unit 7
concrete paver		混凝土地面砖；混凝土铺路机	Unit 10
concrete plant		混凝土厂	Unit 4
concrete ['kɒŋkri:t]	*n.*	混凝土	Unit 2
condenser [kən'densə(r)]	*n.*	冷凝器	Unit 5
configuration [kən,fɪgə'reɪʃn]	*n.*	外形，构造（形式）	Unit 1
conformance [kən'fɔ:məns]	*n.*	一致性；符合性	Unit 4
consolidate [kən'sɒlɪdeɪt]	*v.*	固结，加固	Unit 3
conspicuous [kən'spɪkjuəs]	*adj.*	明显的	Unit 2
constant flow rate controller valve		恒流量控制阀	Unit 9
constituent [kən'stɪtjuənt]	*n.*	构成，组成部分	Unit 1
consultant [kən'sʌltənt]	*n.*	（顾问）工程师，监理工程师	Unit 1
consumer terminal		用户终端	Unit 9
contour ['kɒntʊə(r)]	*n.*	地形，等高线	Unit 2
contraction [kən'trækʃn]	*n.*	收缩，缩减	Unit 4
contractor [kən'træktə(r)]	*n.*	承包商	Unit 1

conveyor [kən'veɪə(r)]	*n.*	输送设备；传送带	Unit 9
corrosion [kə'rəʊʒn]	*n.*	腐蚀，侵蚀，锈蚀	Unit 4
corrugated iron		瓦楞铁，铁皮波纹瓦	Unit 2
corrugated metal		金属波纹瓦；压型钢板	Unit 5
course [kɔː(r)s]	*n.*	流向	Unit 2
courtyard ['kɔ:tjɑ:d]	*n.*	院子	Unit 2
covenant ['kʌvənənt]	*n.*	契约	Unit 2
crack [kræk]	*v.*	断裂，折断	Unit 4
crack ['kræk]	*v.*	开裂	Unit 7
cracking ['krækɪŋ]	*n.*	裂缝，裂纹	Unit 3
crane [kreɪn]	*n.*	吊车，起重机	Unit 4
curing ['kjʊərɪŋ]	*n.*	固化；养护	Unit 4
cut-and-fill		挖填方	Unit 10
cylindrical [sə'lɪndrɪkl]	*adj.*	圆柱形的，圆筒状的	Unit 9

D

damp [dæmp]	*adj.*	潮湿的	Unit 7
day joint		施工缝，后接缝	Unit 4
dead load		恒载；静荷载	Unit 4
deep foundation		深基础	Unit 3
deflection [dɪ' flekʃn]	*n.*	变位，偏移	Unit 4
deformation [ˌdi:fɔ:'meɪʃn]	*n.*	变形	Unit 4
deliver [dɪ'lɪvə(r)]	*v.*	提交	Unit 1
demolish [dɪ'mɒlɪʃ]	*v.*	拆除	Unit 2
demolition [ˌdemə'lɪʃn]	*n.*	拆除	Unit 2
demountable [di:'maʊntəbl]	*adj.*	可拆卸的，可卸下的	Unit 8
designate ['dezɪgneɪt]	*v.*	指明，指出	Unit 4
detail ['di:teɪl]	*n.*	详图，大样	Unit 1
deteriorate [dɪ'tɪərɪəreɪt]	*v.*	变质，老化，退化	Unit 6
deterioration [dɪˌtɪərɪə'reɪʃn]	*n.*	衰退	Unit 8
diagram ['daɪəgræm]	*n.*	（示意）图，计算图表	Unit 1
diaphragm load		隔板荷载	Unit 7
differential movement		不均匀变形	Unit 7
differential settlement		不均匀沉降	Unit 3

digger ['dɪgə(r)]	*n.*	挖掘者，挖掘机	Unit 1
dimensional change		尺寸变化	Unit 6
disperse [dɪ'spɜ:s]	*v.*	（使）分散	Unit 10
dispersion [dis'pɜ:ʃən]	*n.*	驱散；散布	Unit 5
distribution board		配电板，配电盘	Unit 9
domestic building		民用建筑	Unit 5
DPC (Damp Proof Course)		防水（潮）层	Unit 7
DPM (Damp Proof Membrane)		防水膜	Unit 7
drafter ['drɑ:ftə(r)]	*n.*	草图设计员；绘图员	Unit 4
draftsperson ['drɑ:ftspə:sən]	*n.*	绘图员	Unit 1
drain [dreɪn]	*v.*	排水	Unit 3
drainage ['dreɪnɪdʒ]	*n.*	排水	Unit 5
drainage channel		排水沟	Unit 10
drained cavity construction		疏水片导流排水法	Unit 3
drainlayer [drein'leiə]	*n.*	铺设排水管的工人，排水管铺设机	Unit 1
drawing ['drɔ:ɪŋ]	*n.*	图纸	Unit 1
ducting ['dʌktɪŋ]	*n.*	管道，导管	Unit 8
duct-mounted silencer		管式消声器	Unit 9
ductwork ['dʌktˌwɜ:k]	*n.*	管道系统，管道	Unit 9
durability [ˌdjʊərə'bɪlətɪ]	*n.*	耐久性	Unit 5
dwelling ['dwelɪŋ]	*n.*	住宅，寓所	Unit 2

E

earth connection		接地线	Unit 9
earth work		土方工程，土石方工程	Unit 3
easement ['i:zmənt]	*n.*	地役权，供役权	Unit 2
eccentric [ɪk'sentrɪk]	*adj.*	偏心的	Unit 6
eccentric load		偏心荷载	Unit 6
eggcrate ['egkreɪt]	*n.*	蛋形格栅，花格	Unit 8
electrician [ɪˌlek'trɪʃn]	*n.*	电工	Unit 1
electrocution [ɪˌlektrə'kju:ʃn]	*n.*	触电，触电死亡	Unit 9
elevation [ˌelɪ'veɪʃn]	*n.*	立面图	Unit 1
elongating [i:'lɒŋgeɪtɪŋ]	*n.*	延长，加长	Unit 4
encase [ɪn'keɪs]	*v.*	围住，包住	Unit 4

encumbrance [ɪn'kʌmbrəns]	*n.*	限定	Unit 2
end return		端墙（return 迂回墙）	Unit 6
energy efficient		能源高效	Unit 1
engineering [ˌendʒɪ'nɪərɪŋ]	*n.*	工程	Unit 1
equation [ɪ'kweɪʒn]	*n.*	方程式；等式	Unit 4
escalator ['eskəleɪtə(r)]	*n.*	自动扶梯	Unit 9
estate [ɪ'steɪt]	*n.*	房地产	Unit 2
excavate ['ekskəveɪt]	*v.*	挖掘，开挖	Unit 3
excavation [ˌekskə'veɪʃn]	*n.*	挖掘	Unit 2
exhaust fans		排气扇	Unit 5
expansion [ɪk'spænʃn]	*n*	膨胀	Unit 6
external [eks'tɜ:nl]	*adj.*	外面的	Unit 1
external works		室外工程，外部工程	Unit 10
extruded aluminum		挤压铝型材，压制铝材	Unit 8

F

fabric ['fæbrɪk]	*n.*	结构，骨架	Unit 1
fabricate ['fæbrɪkeɪt]	*v.*	制造；组合	Unit 4
fabrication [ˌfæbrɪ'keɪʃn]	*n.*	装配	Unit 2
fall [fɔ:l]	*n.*	坡度，斜度，坡降	Unit 10
false ceiling		假平顶，吊顶	Unit 8
fan strutting		扇形撑	Unit 5
feed tank		给水箱	Unit 9
fill [fɪl]	*n.*	填土，回填料	Unit 7
finish ['fɪnɪʃ]	*n.*	装修	Unit 1
finish ['fɪnɪʃ]	*n.*	外表装饰，表层，面层抹灰	Unit 8
fire damper		防火档板；防火阀；防火风门	Unit 9
fire-stop		阻火材料，阻火部件	Unit 6
fitting ['fɪtɪŋ]	*n.*	配件，家（灯）具，器具	Unit 1
fixed appliance		固定设备	Unit 9
fixings ['fɪksɪŋz]	*n.*	紧固件	Unit 5
flat roof		平屋面	Unit 1
flexible [fleksəbl]	*adj*	柔性的	Unit 5
flexible ducting		柔性管道	Unit 9

floating slab		浮飘板，悬浮板	Unit 7
footing ['fʊtɪŋ]	*n.*	基脚	Unit 2
foreman bricklayer		砌筑工工长，瓦工工长	Unit 5
format ['fɔ:mæt]	*n.*	样式；规格	Unit 8
formwork ['fɔ:mwɜ:k]	*n.*	模板	Unit 2
foundation [faun'deiʃən]	*n.*	地基	Unit 2
framework ['freɪmwɜ:k]	*n.*	框架；构架	Unit 4
frog [frɒg]	*n.*	(砖面)凹槽	Unit 6
frost line		冰冻线，冻深线	Unit 3
furnishing ['fɜ:nɪʃɪŋ]	*n.*	家具；装饰品	Unit 4
fuse [fju:z]	*n.*	保险丝；导火线	Unit 9

G

gang [gæŋ]	*n.*	一组（人）	Unit 10
gas service		燃气设施	Unit 9
General Purpose Outlet		通用插座	Unit 9
glass block		玻璃砖	Unit 6
glazed panel		玻璃板	Unit 8
glazing ['gleɪzɪŋ]	*n.*	门窗格玻璃，装（配）玻璃；镶嵌玻璃	Unit 5
gradient ['greɪdiənt]	*n.*	梯度，倾斜度	Unit 2
gravel ['grævl]	*n.*	砾石，碎石	Unit 10
gravity ['grævəti]	*n.*	地心引力；重力	Unit 4
groove [gru:v]	*n.*	沟，槽	Unit 4
ground floor system		底层结构	Unit 10
ground level	.	地面标高	Unit 10
groundworker ['graʊndwɜ:kə]	*n.*	挖土方的工人，铺路工	Unit 1
grout [graʊt]	*v.*	勾缝，灌浆	Unit 7

H

halve [hɑ:v]	*v.*	半对搭；相嵌接合	Unit 5
hand drier		干手器	Unit 6
handover ['hændəʊvə (r)]	*n.*	移交	Unit 10
hard landscape		硬绿化，硬质景观	Unit 10
hardcore ['hɑ:dkɔ:]	*n.*	碎砖垫层，硬石填料	Unit 7

heating ['hi:tɪŋ]	*n.*	供暖；加热	Unit 4
heaving ['hi:vɪŋ]	*n.*	抬起，隆起	Unit 3
helical ['helɪkl]	*adj.*	螺旋形的，螺纹的	Unit 3
hemp type fibre insulation		麻制类纤维绝缘材料	Unit 9
hold-down clip		固定卡子	Unit 8
hold-up [həʊld ʌp]	*n.*	耽误	Unit 10
honeycomb		蜂窝	Unit 9
hot and cold reticulated water system		冷热水循环系统	Unit 9
hut [hʌt]	*n.*	棚屋，临时用房	Unit 2

I

imitation stone		人造石	Unit 10
impact resistance		抗冲击性	Unit 7
impermeability[ɪmˌpɜ:mɪə'bɪlətɪ]	*n.*	抗渗性	Unit 7
imposed load		活荷载，可变荷载，附加荷载	Unit 5
infill panel		填充板	Unit 8
infill panel		内镶（嵌）板	Unit 6
inhibit [ɪn'hɪbɪt]	*v.*	禁止，抑制	Unit 8
in-situ [ˌɪn'saɪtu:]	*adj.*	现场的；原位的	Unit 4
instruction [ɪn'strʌkʃn]	*n.*	指令	Unit 1
insulation [ˌɪnsju'leɪʃn]	*n.*	绝缘或隔热的材料；隔声	Unit 5
intermediate coat		二道底漆，中层漆	Unit 8
internal [ɪn'tə:nəl]	*adj.*	内部的	Unit 1
intersecting wall		相交墙，交叉墙	Unit 6
intumescent material		发泡型防火材料	Unit 9
irregularity [ɪˌregjə'lærətɪ]	*n.*	不规则；不整齐	Unit 7

J

joint [dʒɔɪnt]	*n.*	接缝，接合处；	Unit 4
junction ['dʒʌŋkʃn]	*n.*	接合点，连接点	Unit 9

L

labeled ['leɪbld]	*adj.*	带有标记的	Unit 1
laitance ['leɪtəns]	*n.*	翻沫；水泥乳	Unit 4

laminate ['læmɪnət]	*adj.*	由薄片叠成的	Unit 8
Lands Titles Office		土地所有权办公室	Unit 2
landscape ['lændskeɪp]	*n.*	园林，景观	Unit 1
Landscaping gardening		园林园艺	Unit 10
lateral ['lætərəl]	*adj.*	侧面的，横向的	Unit 6
lateral movement		侧向变形，侧向移动	Unit 3
layer ['leiə]	*n.*	层，夹层	Unit 3
layout ['leɪaʊt]	*n.*	布置，定位（线），放样	Unit 1
layout drawing		平面布置图，规划图	Unit 10
layout ['leɪaʊt]	*n.*	布局	Unit 2
leanmix ['li: nmɪsk]	*n.*	贫拌合料，少灰混合	Unit 3
letter of acceptance		中标函	Unit 1
level ['levl]	*v.*	拉平；找平	Unit 7
level ['levl]	*n.*	水平线，水平面	Unit 8
licensed plumber		执照管道工，注册管道工	Unit 9
lift pit		电梯井底坑，电梯基坑	Unit 3
lift shaft		电梯井，（升降）机竖井	Unit 3
lighting column		灯柱	Unit 10
lightweight concrete		轻质混凝土	Unit 5
line ['laɪn]	*v.*	给……加内衬，排列，衬砌	Unit 6
lintel ['lɪntl]	*n.*	过梁	Unit 5
litter bin		废物箱，垃圾箱	Unit 10
live load		活荷载	Unit 4
load [ləʊd]	*n.*	荷载	Unit 1
load bearing edge beam		承重边梁	Unit 7
load bearing wall		承重墙	Unit 5
loadbearing [ləud 'bɛəriŋ]	*n.*	承载，承重	Unit 4
lot number		地块（地段）编号，批号	Unit 2
louver ['lu:və]	*n.*	百叶（式），格栅；百叶窗	Unit 6
luminary ['lu:mɪnərɪ]	*n.*	照明灯，发光体	Unit 8

M

main [meɪn]	*n.*	总管道	Unit 2
mains pressure		总管压力	Unit 9

maintenance ['meɪntənəns]	*n*	维修，保养	Unit 6
manufacturer [ˌmænju'fæktʃərə(r)]	*n.*	制造商	Unit 1
mar [mɑː(r)]	*v.*	毁坏	Unit 8
masonry ['meɪsənri]	*n.*	砖石；砌体	Unit 4
masonry wall		砌体墙	Unit 6
mass [mæs]	*n.*	[物理学]质量	Unit 4
mat [mæt]	*n.*	混凝土垫层	Unit 3
mechanical[mɪ'kænɪk(ə)l]	*adj.*	机械的，设备的，力学的	Unit 7
members ['membə(r)]	*n.*	构件	Unit 5
membrane ['membreɪn]	*n.*	膜（片），隔板，防渗护面，表层	Unit 3
membrane wall		夹芯墙	Unit 6
metal etch primer		金属磷化底漆	Unit 8
metal primer		金属底漆	Unit 8
meter ['miːtə(r)]	*n.*	表；计量器	Unit 9
meter box		仪表箱，电表箱	Unit 9
minute cavity		细微孔洞，微小空洞	Unit 8
mixing plant		搅拌厂	Unit 2
modular ['mɒdjələ(r)]	*adj.*	模数，（建筑等）组合式的，模块化的	Unit 8
moisture content		含水量，含水率，湿度	Unit 3
moisture penetration		潮湿的侵袭	Unit 8
moisture ['mɒɪstʃə]	*n.*	潮气，水分，湿度	Unit 7
mold [məʊld]	*n.*	模子；模具	Unit 4
monolithic [ˌmɒnə'lɪθɪk]	*adj.*	整体的；庞大的	Unit 4
monolithic wall		实心墙	Unit 6
mortar ['mɔːtə(r)]	*n.*	砂浆	Unit 6
mortar joint		灰缝，砂浆接缝	Unit 6
mould growth		长霉	Unit 8
municipal [mjuː'nɪsɪpl]	*adj.*	市政的，市的	Unit 2

N

newton ['njuːtən]	*n.*	牛顿（力的单位）	Unit 4
nogging ['nɒgɪŋ]	*n.*	横撑杆件，木砖，木架砖壁（木架中填砖）	Unit 8
non-sewerage area		非排污水区	Unit 9

nuisance ['nju:sns]	*n.*	损害	Unit 10

O

obliterate [ə'blɪtə,reɪtɪŋ]	*v.*	除去，除掉	Unit 8
obstruction [əb'strʌkʃn]	*n.*	障碍物	Unit 2
occupants ['ɒkju:pənts]	*n.*	居住者，占有人	Unit 4
opening ['əʊpnɪŋ]	*n.*	洞口；开口	Unit 6
orientation [,ɔ:riən'teɪʃn]	*n.*	方位	Unit 2
ornamental finish		装饰精修	Unit 10
outcrop ['aʊtkrɒp]	*n.*	砂层，层	Unit 3
outweigh [,aʊt'weɪ]	*v.*	胜过，强过	Unit 7
overflow relief gully		溢流排水沟	Unit 10
overlap [,əʊvə'læp]	*v.*	搭接，重叠	Unit 6
overturn [,əʊvə'tɜ:n]	*v.*	垮，倾覆	Unit 6

P

pad [pæd]	*n.*	独立基础	Unit 1
paint [peɪnt]	*n.*	油漆；涂料	Unit 8
panel ['pænl]	*n.*	板，镶板，嵌板	Unit 6
partition [pɑ:'tɪʃn]	*n.*	隔墙，隔断	Unit 6
party wall		分户墙，共用隔墙，界墙	Unit 5
patchy ['pætʃɪ]	*adj.*	不调和的；拼凑成的	Unit 8
patented connector		专利连接件，特制接头	Unit 9
pave [peɪv]	*v.*	铺设	Unit 10
pavement ['peɪvmənt]	*n.*	路面，硬化区，（园林）小路	Unit 1
paving level		路面标高	Unit 10
pavior ['peɪvjə]	*n.*	铺砌工人，石匠	Unit 1
penetration [,penɪ'treɪʃn]	*n.*	渗透，穿透	Unit 6
performance	*n.*	功能，性能；运行	Unit 5
performance security		履约担保	Unit 1
perimeter [pə'rɪmɪtə(r)]	*n.*	周边，边界	Unit 3
perpendicular [,pɜ:pən'dɪkjələ(r)]	*adj.*	垂直的；垂直	Unit 5
perspective [pə'spektɪv]	*n.*	透视图，远景	Unit 1
pile [paɪl]	*n.*	桩基础	Unit 1

pile cap		桩承台，桩帽	Unit 3
pit [pɪt]	*n.*	坑，取土坑，料坑	Unit 3
pitched roof		坡屋顶	Unit 1
plank [plæŋk]	*n.*	板，支持物；厚木板	Unit 7
plant [plɑ:nt]	*n.*	厂；工厂	Unit 4
plasterer ['plɑ:stərə(r)]	*n.*	抹灰工	Unit 1
plastic soil		塑性土壤，可塑土	Unit 3
platform ['plætfɔ:m]	*n*	平台	Unit 2
plumber ['plʌmə(r)]	*n.*	管工，水暖工	Unit 1
plumbing ['plʌmɪŋ]	*n.*	管道；给排水系统，管件安装	Unit 4
plumbing fix		管道安装	Unit 9
polystyrene [ˌpɒlɪ'staɪri:n]	*n.*	聚苯乙烯	Unit 8
polythene sheeting		聚乙烯护板	Unit 7
ponding ['pɒndɪŋ]	*n.*	积水	Unit 5
porch [pɔ:tʃ]	*n.*	走廊，门廊	Unit 7
porosity [pɔ:'rɒsətɪ]	*n.*	孔隙度；孔隙率	Unit 8
porous ['pɔ:rəs]	*adj.*	多孔的，能渗透的	Unit 6
post-tension slab		后张法预应力板	Unit 3
pour [pɔ:(r)]	*v.*	灌，浇筑（混凝土）	Unit 3
pouring [pɔ:rɪŋ]	*n.*	浇筑	Unit 7
precast [ˌpri:'kɑ:st]	*adj.*	预浇筑的，预制的	Unit 4
precast concrete		预制混凝土	Unit 4
precision-made rigid plastic pipe		精密制造的硬质塑料管	Unit 9
premature failing		过早损坏	Unit 8
premium ['pri:mɪəm]	*n.*	保险费	Unit 1
pressed steel		压制型钢，冲压型钢	Unit 8
pressure limiting valve		限压阀	Unit 9
prestressed [ˌpri:'strest]	*adj.*	预应力的	Unit 4
pre-tensioned concrete		先张法（预应力）混凝土	Unit 3
primer ['praɪmə(r)]	*n.*	(用于基层保护或与底漆结合的）打底漆	Unit 8
professional [prə'feʃənl]	*adj.*	专业的	Unit 1
profiled aluminum tray		预制铝合金槽	Unit 8
protective coat		保护涂层	Unit 8

provision [prə'vɪʒn]	*n.*	装置，设备，临时设施	Unit 2
puncture ['pʌŋktʃə(r)]	*v.*	刺穿	Unit 7
puncture resistance		抗穿刺性，抗冲击	Unit 7
purlin ['pɜ:lɪn]	*n.*	檩条	Unit 5
putty ['pʌtɪ]	*n.*	腻子；堵料	Unit 8

Q

quality control		质量控制	Unit 1
quantity surveyor		估算师，计量员	Unit 1

R

raft [rɑ:ft]	*n.*	筏形基础，排基	Unit 1
raft slab		筏板，筏形基础	Unit 7
rafter ['rɑ:ftə(r)]	*n.*	椽条；椽子	Unit 5
ramp [ræmp]	*n.*	坡道	Unit 10
reactive [rɪ'æktɪv]	*adj.*	活性的	Unit 3
rectify ['rektɪfaɪ]	*v.*	调整；矫正	Unit 8
reduced pressure		减压	Unit 9
refer [rɪ'fɜ:(r)]	*v.*	送交	Unit 6
refuse [rɪ'fju:z]	*n.*	垃圾，废弃物	Unit 9
reinforced concrete		钢筋混凝土	Unit 6
reinforced mat slab		钢筋混凝土筏板或平板	Unit 3
reinforcement [ˌri:ɪn'fɔ:smənt]	*n.*	钢筋，加固	Unit 2
releasing agent		防粘剂，释放剂	Unit 4
rendering ['rendərɪŋ]	*n.*	罩面砂浆，初涂，打底	Unit 8
resident architect	*n.*	驻现场建筑师	Unit 1
resistance [rɪ'zɪstəns]	*n.*	抵抗；阻力；抗力；电阻	Unit 5
restraint [rɪ'streɪnt]	*n.*	固定，约束，抑制	Unit 8
retaining wall		挡土墙	Unit 2
ridge [rɪdʒ]	*n.*	屋脊	Unit 5
rigidity [rɪ'dʒɪdətɪ]	*n.*	刚性；刚度	Unit 4
rolled section		辊压断面	Unit 4
rod [rɒd]	*n.*	杆，拉杆，钢筋	Unit 4
roof cladding		屋顶覆盖材料；屋面覆盖层	Unit 5

roof strutting		屋顶支撑	Unit 5
roof trusses		屋架	Unit 5
rotation [rəʊ'teɪʃn]	*n.*	旋转，扭转	Unit 6
rugged ['rʌgɪd]	*adj.*	粗糙的	Unit 10
run [rʌn]	*n.*	管线布局	Unit 8
runner ['rʌnə(r)]	*n.*	（吊顶）横龙骨，龙骨，中间支承干	Unit 8
rust [rʌst]	*v.*	生锈	Unit 8

S

sagging ['sægɪŋ]	*n.*	下沉	Unit 5
sand [sænd]	*n.*	砂	Unit 3
sanitary ['sænɪtrɪ]	*adj.*	卫生的，清洁的	Unit 2
sanitation [ˌsænɪ'teɪʃn]	*n.*	卫生设备（排水设备），下水道设备	Unit 1
scaffold ['skæfəʊld]	*n.*	脚手架	Unit 2
scaffolder	*n.*	架子工	Unit 1
screed [skri:d]	*n.*	砂浆底层，找平层	Unit 7
sealer ['si:lə(r)]	*n.*	密封剂，封底漆	Unit 8
second fix		二次安装	Unit 9
section ['sekʃn]	*n.*	剖面图	Unit 1
steel section		型钢	Unit 5
semi-dry method		半干法	Unit 7
separating layer		分离层	Unit 7
separating wall		隔墙，分户墙	Unit 6
septic and effluent waste		污水废水	Unit 9
septic tank		化粪池	Unit 9
service ['sɜ:vɪs]	*n.*	设备	Unit 1
service tunnel		地下管道	Unit 3
set out		放样	Unit 2
settlement ['setlmənt]	*n.*	沉降，建筑物的下沉	Unit 3
sewage disposal		污水处理	Unit 9
sewage ['su:ɪdʒ]	*n.*	下水道	Unit 2
sewerage plumbing		污水管道工程，下水道系统	Unit 9
sewerage system		污水系统，污水排放系统	Unit 9
shaft [ʃɑ:ft]	*n.*	（竖，升降，通风）井	Unit 3

shallow foundation		浅基础	Unit 3
shear [ʃɪə(r)]	*v.*	切变；切断	Unit 4
shear failure		剪切断裂，剪切破坏	Unit 3
shear force		剪力	Unit 4
shop drawing		制造图，施工图，深化设计图	Unit 4
shrinkage ['ʃrɪŋkɪdʒ]	*n.*	收缩	Unit 6
shrinking [ʃrɪŋkɪŋ]	*n.*	收缩	Unit 3
shuttering ['ʃʌtərɪŋ]	*n.*	模板 [壳]	Unit 4
site agent		现场经理，项目经理	Unit 1
site drainage plan		施工现场排水平面图	Unit 10
site engineer		工地工程师，施工员	Unit 2
site investigation		现场调查，场地勘察	Unit 10
site office		工地办公室	Unit 2
site plan		现场（场地）平面图	Unit 2
skeletal ['skelətl]	*n.*	框架的，骨架的	Unit 1
sketch [sketʃ]	*n.*	草图，设计图	Unit 1
skim coat		罩面层；薄覆盖层	Unit 8
skylight ['skaɪlaɪt]	*n.*	天窗	Unit 1
slab [slæb]	*n.*	板	Unit 3
slab-on-grade [slæb ɒn greɪd]	*n.*	地面（混凝土）板，底板式基础	Unit 3
slate [sleɪt]	*n.*	石板瓦，石板	Unit 5
slenderness ['slendənɪs]	*n.*	长细比，细长度	Unit 6
slide [slaɪd]	*v.*	滑移，滑落；错动	Unit 4
slope [sləʊp]	*n.*	斜坡； 斜面	Unit 10
soakage ['səʊkɪdʒ]	*n.*	浸润，浸透，浸渍	Unit 8
socket ['sɒkɪt]	*n.*	插座；灯座	Unit 9
soffit ['sɒfɪt]	*n.*	拱腹；下表面；底面	Unit 8
soft landscape		软绿化，软质景观	Unit 10
sound absorber		吸声材料	Unit 6
sound reduction		隔音	Unit 6
span [spæn]	*n.*	跨度	Unit 4
Special Purpose Outlet		专用插座	Unit 9
specification [ˌspesɪfɪ'keɪʃn]	*n.*	规范；技术说明书	Unit 1
specify ['spesɪfaɪ]	*vt.*	说明，明确提出	Unit 10

spirit level		水平尺，水平仪	Unit 8
spoon drain		V 形排水沟；匙状排水沟	Unit 10
spread footing		大放脚，扩展基础	Unit 3
spreader cleat		加固托座	Unit 5
sprinkler ['sprɪŋklə(r)]	*n.*	（建筑物内的）自动喷水灭火装置	Unit 9
stabilized [s'teɪbəlaɪzd]	*adj.*	稳定的	Unit 3
stanchion ['stæntʃən]	*n.*	柱子，标柱，标桩	Unit 3
statutory ['stætjʊtrɪ]	*adj.*	法定的，法令的	Unit 9
steel decking		（压型）钢板，钢衬板；钢板层	Unit 7
steel section		型钢	Unit 5
step ladder		人字梯，登高梯	Unit 8
stiff [stɪf]	*adj.*	刚性的	Unit 5
stiffening ['stɪfnɪŋ]	*n.*	加强，加劲	Unit 4
stockpile ['stɒkpaɪl]	*v.*	堆放	Unit 2
stockpiling [s'tɒkpaɪlɪŋ]	*n.*	贮存；堆放	Unit 4
storey ['stɔ:rɪ]	*n.*	楼层	Unit 2
strain [streɪn]	*n.*	应变；张力	Unit 4
strength [streŋθ]	*n.*	强度	Unit 5
stress [stres]	*n.*	应力	Unit 4
stretcher bond		顺砖砌合，全顺组砌法	Unit 6
stretching ['stretʃɪŋ]	*n.*	拉伸，伸长	Unit 4
strip [strɪp]	*n.*	条形基础；长条，板条	Unit 1
strip [strɪp]	*v.*	除去，剥去；拆模	Unit 7
strip footing		条形基础	Unit 3
strip foundation		条形基础；条形地基	Unit 4
structural ['strʌktʃərəl]	*adj.*	结构的	Unit 1
structural failure		结构损坏，结构性破坏，结构失效	Unit 3
structure ['strʌktʃə(r)]	*n.*	结构，构筑物	Unit 1
strut [strʌt]	*n.*	对角撑；撑杆	Unit 5
strutting beams		支撑梁	Unit 5
stud [stʌd]	*n.*	龙骨；立筋	Unit 5
stud wall		立筋隔墙；立柱墙	Unit 5
sub-circuit		子电路，支回路	Unit 9
subcontractor [ˌsʌbkən'træktə(r)]	*n.*	分包商	Unit 1

subdivide ['sʌbdɪvaɪd]	*v.*	再分，细分	Unit 8
subsoil ['sʌbsɔɪl]	*n.*	下层土，老土	Unit 3
subsoil drain		地下排水管（道），地下排水沟	Unit 10
substrate ['sʌbstreɪt]	*n.*	基层，基质	Unit 8
sump [sʌmp]	*n.*	集水井，排水沟	Unit 3
superimposed [sju:pərɪm'pəʊzd]	*adj.*	附加的，叠加的	Unit 6
supply authority mains		电源供应局	Unit 9
supporting structure	*v.*	支承结构；固定架	Unit 5
supervise		监督，管理，指导	Unit 5
surface ['sɜ:fɪs]	*adj.*	表层的	Unit 1
surface ['sɜ:fɪs]	*n.*	表面	Unit 8
surface drainage		地面排水	Unit 10
survey ['sɜ:veɪ]	*v.*	测量	Unit 1
survey peg		测量标桩	Unit 2
suspended [sə'spendɪd]	*adj.*	悬挂的，悬挑的，悬空的	Unit 1
suspended ceiling		吊顶，顶棚	Unit 5
suspension hanger		悬挂吊架	Unit 8
swelling ['swelɪŋ]	*n.*	湿胀，膨胀，隆起	Unit 3
switchboard ['swɪtʃbɔ:d]	*n.*	配电盘，控制板	Unit 9
synthetic foam insulation		合成泡沫绝缘材料	Unit 9

T

take over		接收	Unit 1
tear strength		撕裂强度	Unit 7
tensile force		拉力	Unit 4
tensile stress		拉应力	Unit 4
tension ['tenʃn]	*n.*	张力，拉力	Unit 4
termite ['tɜ:maɪt]	*n.*	白蚁	Unit 7
termite barrier provision		白蚁阻隔层规定	Unit 10
thermal ['θɜ:məl]	*adj*	热的，保热的； 温热的	Unit 5
thermal ['θɜ:məl]	*n.*	冷热，温度	Unit 9
thickness ['θɪknəs]	*n.*	厚度	Unit 8
threaded joint		丝扣接头；螺纹接头	Unit 9
tile [tail]	*n.*	瓷砖；面砖，瓦	Unit 7

tiler ['taɪlə(r)]	*n.*	瓦工	Unit 5
timber ['tɪmbə(r)]	*n.*	木材	Unit 1
topography [tə'pɒgrəfɪ]	*n.*	地势，地貌	Unit 2
topping ['tɒpɪŋ]	*n.*	面层，覆盖层	Unit 7
topsoil ['tɒpsɔɪl]	*n.*	表层土	Unit 2
towercrane		塔式起重机	Unit 2
trade [treɪd]	*n.*	工种，行业，专业	Unit 1
trap [træp]	*n.*	存水弯，水封	Unit 9
trench [trentʃ]	*n.*	渠，壕，沟（槽，渠）	Unit 3
truss [trʌs]	*n.*	衍架	Unit 5

U

undercoat ['ʌndəkəʊt]	*n.*	面漆下涂层；底涂层	Unit 8
unearth [ʌn'ɜ:θ]	*n.*	发掘，掘出	Unit 3
uniform ['ju:nɪfɔ:m]	*adj.*	均匀的，匀质的	Unit 3
uninhabitable [ˌʌnɪn'hæbɪtəbl]	*adj.*	不适于居住的	Unit 4
uplift ['ʌplɪft]	*n.*	拔起，提起，隆起	Unit 6
utility service		公用服务设施	Unit 9

V

vapour barrier		隔汽层	Unit 7
vegetation [vedʒɪ'teɪʃ(ə)n]	*n.*	植被	Unit 7
ventilation [ˌventɪ'leɪʃn]	*n.*	通风设备；空气流通	Unit 9
verandah [və'rændə]	*n.*	走廊	Unit 7
vertical ['vɜ:tɪkl]	*adj.*	垂直的	Unit 3
voltage ['vəʊltɪdʒ]	*n.*	电压，伏特数	Unit 9

W

wallboard primer		墙板底漆	Unit 8
water course		水流，水系；水路	Unit 2
water reticulation system		水网系统	Unit 9
water tank		水箱	Unit 5
waterproof ['wɔ:təpru:f]	*adj*	防水的；用防水材料处理过的	Unit 5
waterproofing ['wɔ:təpru:fɪŋ]	*n.*	防水（层），防水作业	Unit 3

watertight ['wɔ:tətaɪt]	*adj.*	水密的，不漏水的	Unit 2
wear and tear		磨损，损坏，损耗	Unit 7
wearing surface		耐磨面，磨耗面，磨损面	Unit 7
weathering ['weðərɪŋ]	*n.*	风化	Unit 5
welding [weldɪŋ]	*n.*	焊接法，定位焊接	Unit 4
wind bracing		抗风支撑；防风拉筋	Unit 4
wind loads		风荷载	Unit 5
wood wool slab		刨花板	Unit 5
woodwool ['wʊdwʊl]	*n.*	木丝，木刨花	Unit 8

Z

zoning ['zəʊnɪŋ]	*n.*	土地（功能）区划，分区制	Unit 2